RAND

Organizing, Training, and Equipping the Air Force for Crises and Lesser Conflicts

Carl H. Builder,
Theodore W. Karasik

Prepared for the
United States Air Force

Project AIR FORCE

Operations short of war, one of the responsibilities of the U.S. armed services, are increasingly consuming the attention and resources of U.S. military forces around the world. The locations of such missions are prominent in the headlines: Bosnia, Somalia, Haiti, Rwanda, and Kuwait. In a few instances and for some capabilities, those burdens appear to have stressed the forces to exhaustion or failure. A public debate has been joined as to whether such assignments are a "proper" use of U.S. military power or whether they are the "wave of the future." The research reported here explores where, why, and how operations short of war are stressing the forces, particularly the aerospace forces, and how those stresses upon USAF capabilities might be relieved by changes in Air Force organization, training, and equipment. The research relies, in part, upon visits to the staffs of all of the regional commanders in chief and their air component commanders, undertaken during the first quarter of 1994.

The research supporting this report was conducted as part of the Crises and Lesser Conflicts (CALCs) project under the Strategy, Doctrine, and Force Structure Program of RAND's Project AIR FORCE. It should be of interest to those Air Force and other U.S. military personnel, analysts, policymakers, and operational commanders who are concerned with the future applications of aerospace power to operations short of war.

PROJECT AIR FORCE

Project AIR FORCE, a division of RAND, is the Air Force federally funded research and development center (FFRDC) for studies and analyses. It provides the Air Force with independent analyses of policy alternatives affecting the development, employment, combat readiness, and support of current and future aerospace forces. Research is being performed in three programs: Strategy, Doctrine, and Force Structure; Force Modernization and Employment; and Resource Management and System Acquisition.

CONTENTS

FIGURE

TABLES

CONSTANT RESPONSIBILITIES, SHIFTING DEMANDS

According to its basic doctrine, "The Air Force is responsible for the preparation of the air forces necessary for the effective prosecution of war and military operations short of war . . . and . . . for the expansion of the peacetime components of the Air Force to meet the needs of war." Preparation of the necessary air forces means to "organize, train, equip, and provide forces" to carry out all the operations required to fulfill the Air Force's primary and collateral functions (now, more commonly called "missions").

Throughout the Cold War, the forces "necessary for the effective prosecution of war" clearly dominated over the other two responsibilities. The Cold War threats required *ready* forces that were capable, if necessary, of prosecuting a war to termination in hours or days. The immediacy and high stakes of those threats made mobilization for war and operations short of war lesser considerations. So, for more than 40 years, the efforts to "organize, train, equip, and provide forces" focused on "the effective prosecution of war," while operations and mobilization short of war were handled as issues on the margins of Air Force priorities. That Cold War focus may be contrasted to the peacetime, pre-World War II era when the emphasis was on mobilization and operations short of war.

With the end of the Cold War, the threat has changed dramatically, but the focus has not. The prospect of war has changed from an imminent collision of nuclear superpowers to what are currently called major regional contingencies (MRCs). In U.S. defense planning, it is argued that the United States must be prepared for two such MRCs at any one time. Accordingly, two simultaneous MRCs have replaced the superpower conflict of the Cold War as the kind of war the Air Force prepares to prosecute when it organizes, trains, equips, and provides its forces—and, once again, operations and mobilization short of war are being handled as issues on the margins, as they were during the Cold War.

Unfortunately, other changes besides the threat of war have accompanied the end of the Cold War: Defense spending has declined steadily as a portion of federal expenditures; and operations short of war have created rising demands for using the U.S. military to solve problems of ethnic conflict, humanitarian and disaster assistance, and civil unrest. The prospect for the remainder of this decade is a continuation of both these trends—fewer resources for the military and more demands for their use in operations short of war—even as the mainstream of U.S. defense planning tries to focus on preparedness for two MRCs.

STRESSING OPERATIONS

Signs of stress are already in evidence on the U.S. military forces being employed for operations short of war:

- Forces and headquarters staffs are stretched thin as they try to handle concurrent or successive commitments to operations short of war.

- Certain critical units are faced with long overseas deployments.

- Units needed for MRCs are committed to operations short of war where they can not be extracted or recovered quickly.

- Training time is being lost, and equipment needed for MRCs is being worn out prematurely.

- Forces are being used in operations for which they were not specifically organized, trained, or equipped.

- Many capabilities most needed for operations short of war are located in the Reserves or National Guards where they may not be available unless a national emergency is declared.

- Clear-cut, fixed military objectives that are both needed and expected for the effective prosecution of war are often absent.

A public debate has been joined over the increasing assignment of the military to these operations short of war. One side argues that the primary purpose of the military is (or should be) to fight and win the nation's wars, while the other side argues that these operations seem to be the wave of the future and the military is the best national resource to address them.

Operations short of war cover a broad spectrum of activities, from domestic calls for disaster assistance and restoration of civil order to international calls for humanitarian aid and military intervention. Some events are relatively short-term, such as relief operations in Bangladesh, relief and evacuation during the Mount Pinatubo disaster, or the evacuations of endangered U.S. citizens. Others are planned as relatively long-term operations, such as the congressionally mandated surveillance of maritime and aerial drug smuggling and the United Nations mandated peacekeeping operations in the Sinai. Still others are of indeterminate length, such as the protection of ethnic minorities in Iraq and Bosnia, awaiting the resolution of political issues.

The particular operations short of war that are now stressing the U.S. military forces are not the domestic or routine operations; they are the *international* and *nonroutine* operations short of war, particularly those that could lead to combat operations that do not develop into MRCs. To distinguish this stressing subset from all other operations short of war and to give these nonroutine, international operations a more precise name, we have chosen to call them crises and lesser conflicts (CALCs). Although crises and lesser conflicts are quite different things, together they make up the entire category of international, nonroutine operations short of war. Just as MRCs have become the dominant form of war for defense planning purposes, CALCs may become the dominant form of operations short of war for defense planning.

THE HORNS OF A DILEMMA

CALCs and MRCs are emerging as two horns of a dilemma for the U.S. military. If the future is dominated by MRCs—actual or threatened—the military will generally have the right kinds of forces, but probably not enough of them. This outcome would not be surprising, because the forces emerging from the Cold War were designed for war; however, as those forces are reduced through budget contractions, the concerns about their adequacy for war are quantitative more than qualitative. On the other hand, if the future is dominated by CALCs, the U.S. military may have enough resources in the aggregate, but not necessarily the right kinds of forces. Thus, the CALC horn presents problems mostly for the *qualities* designed into the U.S. forces, whereas the MRC horn poses problems mostly for retaining sufficient *quantities*. With declining or even constant budgets, efforts to avoid one horn will only increase the problems associated with the other.

Until recently, this dilemma has been masked: During the Cold War, with larger forces and fewer CALCs, the capacities built into the supporting forces were generally adequate to meet the needs. Defense planning proceeded on the assumption that CALCs could be treated as "lesser included cases" for forces designed to handle one or more wars that might emerge if the Cold War turned hot. Some doubted the validity of that assumption for certain kinds of CALCs, even during the Cold War; but the drawdown of the forces—both combat and support—and the increasing numbers of CALCs have laid the horns bare and challenged the assumption that CALCs can continue to be treated as lesser included cases for MRC-designed forces.

AIR FORCE AT THE CUTTING EDGE

Among the U.S. military services, the Air Force is encountering the dilemma sooner and more severely than the other services because many of its unique capabilities are in demand and are already stretched thin by simultaneous CALCs. These stressed capabilities include

- airlift, both global and theater, but especially theater, for the delivery of relief supplies and for the deployment and support of forces;

- surveillance, from air and space, especially airborne warning and control systems (AWACS) for the enforcement of air security;

- reconnaissance and intelligence, from air and space, for situation and risk assessments; and

- ground-to-air threat suppression, such as Wild Weasels, for the enforcement of air security.

At the same time, the bulk of the Air Force's MRC fighting forces—the generic (unspecialized) fighters and bombers—are not particularly stressed, for although they are often required for CALCs, they are not required in the depth or numbers needed for MRCs. Thus, while some of the Air Force assets are being stressed by CALCs, others are not, because the forces have been designed and balanced for MRCs.

AN AIR FORCE DESIGNED FOR CALCS

How much different would the forces be if they had, instead, been designed and balanced for CALCs rather than MRCs? The question is both academic and hypothetical, but that does not mean the answer is without utility as a planning reference point. We know how the Air Force should organize, train, and equip its forces if MRCs are the basis for planning; we do not know, if CALCs were the basis. What we are interested in determining are the *differences* that would result from the two focal points.

Organizing for CALCs: Compared with changes in training or equipment, organizational changes could offer the Air Force some of the least-cost, highest-payoff responses to the problems posed by CALCs. But organizational changes are likely to be the most disturbing to the Air Force as an institution and, therefore, the most difficult to effect.

If CALCs were the principal basis for designing the forces, the biggest organizational change could be in the division of the basic Air Force responsibilities between the active and the Reserve/Guard forces: The prosecution of war could become, as it has before in peacetime, the principal responsibility of the Reserve and Guard units; and thus their mobilization, plus operations short of war, could once again dominate the design of the active force. The active units could be configured for deployments in smaller groupings to meet the needs of multiple, geographically dispersed CALCs, as opposed to being concentrated for MRCs. Crew ratios could be increased where necessary to rotate deployed crews.

A few units might be configured specifically for frequent CALC operations, such as those required for providing security from the air (e.g., Operations Deny Flight and Southern Watch) or air deliveries into unsecured bases (as in the first flights into Somalia). An Air Force designed for CALCs could probably forge more intimate and sustained ties with other organizations—not only with relief and humanitarian nongovernmental organizations (e.g., the International Red Cross), but also with pertinent governmental organizations (e.g., the Nuclear Emergency Search Teams)—whose specialized knowledge or skills may be essential for CALCs.

One of the most important changes in organization for the Air Force to consider is the creation of a headquarters point of advocacy for CALC capabilities. Currently, there are no eyes, ears, or voice within the Air Force to watch, to listen, or to speak for opportunities to improve USAF CALC capabilities, even where those improvements could be achieved at modest costs or with modest changes on the margins—not just through changes in organization, but also in training and equipment. Without a point of advocacy to take the lead in exploring, evaluating, and promoting opportunities to improve USAF CALC capabilities, few of the other suggestions or proposals offered in this report could be expected to find fertile ground or a chance to grow.

Training for CALCs: CALCs may demand education more than training. The difference is more than semantic. Training tends to evaporate and needs frequent refreshing, but education is generally more durable. Most flight and fighting skills are gained and maintained through practice; but CALC duties are more likely to alter how these military skills are to be applied than to introduce entirely new skills. Fliers and warriors believe that they must train continually if they are to keep their flying and fighting skills honed to a sharp edge. By contrast, policemen, once trained at their academies, are likely to limit their training to physical fitness and occasional sessions at the pistol range. Much of their time spent at the police academies is in education—the law, human relations, etc.—subjects that, once learned, will be slowly mastered in the field through long experience rather than by continual training. When fliers or warriors must perform policelike (constabulary) functions in CALCs, their need, like that of the policeman, is more for education than for training.

Most of the skills required for CALCs are already found in the Air Force , but they might be needed in different proportions. Flying and fighting skills are needed for CALCs, but in less depth or numbers than for MRCs. Other skills, such as languages, cultural understanding, and ground security could be needed in greater numbers or depth. It seems likely that many of the skills now found in Special Operations Forces and in the Air Police could be needed in greater numbers for CALCs. The emphasis in some training and exercises could probably shift: Operational training for air security operations might, for example, emphasize rules of engagement and international interoperability that may be more critical for success in CALCs than are sortie generation or air campaign planning. In general, however, many CALC skills could take the form of one-time education or training with occasional refreshers or exercises rather than the steady regimen of training now associated with flying and combat. Like first aid or CPR, simply knowing how to do (or not do) certain things may be sufficient for many CALCs.

Equipping for CALCs: Quantitatively, the equipment of an Air Force designed for CALCs could differ from that of an Air Force designed for MRCs, mostly in its proportions. The numbers of transports, surveillance units, C3I, gunships, and defensive assets in the active force could all increase at the expense of reductions in the numbers of fighters and bombers. The appropriate proportions of equipment could be revealed by the evolution of CALCs, but an examination of recent CALC operations over Bosnia, Iraq, Somalia, and Rwanda should provide a reasonable starting point.

Beyond proportional changes in current Air Force assets, CALC operations could also benefit from some special equipment that will not otherwise be found in an Air Force for MRCs: This is an area in which USAF CALC capabilities might be considerably expanded, albeit at considerable cost. The major options for *qualitatively* improving Air Force equipment for CALCs include abilities to do the following from the air:

* Detect, locate, and immediately suppress heavy weapons fire

* Suppress open urban disorders, without resort to lethal means

- Drop or deliver supplies with PGM (precision guided munitions) accuracy, without landing

- Unload and pick up on short notice a small team of people in any cleared area anywhere in the world, at any time, in any weather

- Deliver large quantities of inexpensive, lightweight, self-erectable, disposable housing and medical structures

- Locate nuclear materials on the ground, at least to the extent now possible with civilian aircraft.

NEW PRIORITIES

Thinking about an Air Force designed for CALCs is only a hypothetical reference point for thinking about what, if anything, should be done now. The most urgent aspect of CALCs in the real world is relieving some of the stresses now falling on certain people, units, and equipment. Unrelieved, these stresses are causing critical people (along with their skills) to leave the Air Force, causing premature wear on critical equipment needed to prosecute both MRCs and CALCs, and setting up the Air Force for failure one way or another. Resources need to be redistributed insofar as possible between fighting and supporting units and between active and Reserve units to relieve these stresses.

Beyond these measures, the Air Force needs to determine how it should respond in the future as the planning world evolves into one dominated by either MRCs or CALCs. That difference in direction now lies on a knife-edge: A single MRC could strongly reinforce the current planning paradigm that focuses on MRCs. But a reunification of the Korean Peninsula and a dramatic change in Persian Gulf regimes could make MRCs seem much more remote if the United States is heavily committed in CALCs. What the Air Force needs right now in CALC planning is not more money but more thought. A starting point would be to recognize that currently there are

- no institutional or bureaucratic pressures on the Air Force to realign its capabilities toward CALCs at the expense of those for MRCs, and

- no points of advocacy within the Air Force for CALC planning concepts—perhaps because such advocacy could be perceived as having the potential to create additional pressures on scarce resources.

Fears of budget pressures, however, should not prevent the Air Force from thinking now about what kinds of actions would be prudent if CALCs should continue to grow in number and scope and then dominate the Air Force operations of the future.

A COMPASS FOR THE FUTURE

The concerns that now impede thinking about CALCs as an important aspect of the Air Force's future include limited resources and problems related to the Air Force's current organization and established traditions. If these considerable concerns can somehow be overcome, what then? What concepts, strategies or doctrines should guide the Air Force as it proceeds to organize, train, and equip forces for CALCs as well as for MRCs? One principle stands out from this research: CALCs can be quagmires. If air power is to offer a significant military alternative for the nation's leadership, it must be allowed to independently carry out activities required in CALCs without committing people to the ground, even in supporting roles. This is not the traditional call for the independence of air power from ground commanders; it is a call for air power to give the nation's leadership an alternative that does not make the nation a hostage in someone else's conflict. Air power must be able to feed, supply, rescue, police, and punish from the air, without resort to air bases within the afflicted area.

This challenge for air power is less technical than financial, and it is less financial than institutional: If the institutional Air Force makes up its mind to pursue such independent capabilities for air power in CALCs, the resources will be found. And if the resources are found, even in an era of sharply constrained budgets, the technical problems can be solved.

This challenge for air power should not be unfamiliar. It is closely related to the challenges for air power that arose in the aftermaths of the two world wars. After World War I, the challenge for air power was to offer the nation's leadership a military alternative to the stalemated carnage of trench warfare on the ground. Air power offered the promise of leaping over those trenches and striking at the heart of the enemy—and avoiding the bloody ground warfare that had cost the Europeans a generation of young men.

After World War II, the challenge for air power was to offer the nation's leadership a military alternative to ground warfare against hordes of soldiers that the United States could not hope to match in numbers. Air power, then pumped up with nuclear weapons, again offered the promise of leaping over the masses of soldiers and striking at the heart of the enemy—and avoiding the kind of attrition warfare on the ground that the nation could not hope to win.

The pattern is evident: After each world war, air power developed by responding to the challenge posed not so much by the *next* war as by the nation's *nightmare* evoked by the last war. Today, the nation's nightmare does not seem to be an MRC, which may be the U.S. military's standard for a "proper" war that can be fought and won. Rather, the nation's nightmare seems to be about finding itself held hostage—as it was in Vietnam, as the Soviets were in Afghanistan—in an endless, unwinnable conflict.

Now, after the Cold War, the challenge for air power could very well be to offer the nation's leadership military alternatives to crises and lesser conflicts that the nation wants neither to ignore nor to be held hostage by. Air power—with independent capabilities to feed, supply, rescue, police, and punish from the air—could be fashioned to address urgent problems without being held hostage.

The challenge is there. So are the means, both technical and financial. But the challenge may not seem worthy of the costs—costs now measured mostly in what the institution has come to value—in traditional forces. The future development and evolution of air power could be in the balance. It has been before, in the 1930s, when the Army leadership thought that air power should be a service rather than a force. It was once again, in the late 1940s, when the Army and Navy leaderships thought that air power should not be independent from their surface forces. And it may be now, over the relevance of air power to a world in which regular warfare seems less likely than the disorders and human tragedies that are increasingly emerging everywhere.

The authors are indebted in this research to their colleagues, whose intellectual contributions to our thinking have been so extensive that we can no longer separate our ideas from theirs. Thus, we gratefully acknowledge the participation in this research of Jack Craigie, Steve Hosmer, Dana Johnson, and Joe Kechichian of RAND; of Colonel Chuck Gagnon, Lieutenant Colonel Mark Anderson, Lieutenant Colonel Ed Wright, and Major George Gagnon of the U.S. Air Force; and Lieutenant Commander Fred Smith of the U.S. Navy. Among these, we should give special credit to Colonel Chuck Gagnon, who gave the research early and vigorous support from the Air Staff, and to Jack Craigie, who coordinated our tightly scheduled visits to the staffs of the regional commanders and their air component commanders.

This report has benefited from critical reviews and constructive comments by Lieutenant Colonel Mark Anderson, Colonel John Warden, and RAND colleague Milt Weiner. Both Warden and Weiner are pioneer thinkers about the application of airpower to unconventional or nontraditional military missions. Of course the authors—and not our colleagues or reviewers—must be held responsible for the inevitable errors and omissions that have escaped into print to haunt us.

AFB	Air Force base
AF/XOX	Directorate of Plans, Deputy Chief of Staff (Plans and Operations), Headquarters, United States Air Force
AFSOF	Air Force Special Operations Forces
AMC	Air Mobility Command
AOR	Area of responsibility
APADS	Advanced precision airborne delivery system
AWACS	Airborne warning and control system
C^3I	Command, control, communications, and intelligence
CALC	Crises and lesser conflicts
CinC	Commander in Chief
COMAIRSOUTH	Commander, Allied Air Forces, Southern Europe
CPR	Cardiopulmonary resuscitation
CRS	Congressional Research Service
DOE	United States Department of Energy
ECM	Electronic countermeasure
EMP	Electromagnetic pulse
FFRDC	Federally funded research and development center
GAO	United States General Accounting Office
GPS	Global positioning system
GTMO	Guantanamo Bay, Cuba (U.S. Naval base at)
JCS	Joint Chiefs of Staff
JTF	Joint task force

LIC	Low-intensity conflict
LRC	Lesser regional conflict
LRCA	Long-range combat aircraft
MAC	Military Airlift Command
MAD	Magnetic anomaly detection
MASH	Mobile Army surgical hospital
MOOTW	Military operations other than war
MRC	Major regional contingency
MRE	Meals ready to eat
NEST	Nuclear emergency search team
NGO	Nongovernmental organization
NLW	Nonlethal weapon
OOTW	Operations other than war
OPTEMPO	Tempo of operations
PGM	Precision guided munition
PME	Professional military education
PSYOPS	Psychological operations
RAF	Royal Air Force (British)
RED HORSE	Rapid engineer deployable, heavy operational repair squadron teams
ROE	Rules of engagement
SEAD	Suppression of enemy air defense
SEAL	Sea, air, and land (Navy teams)
SOCRATES	Special operations command research, analysis, and threat evaluation system
SOF	Special Operations Forces
SOUTHAF (FWD)	Southern Air Forces (forward deployed)
UN	United Nations
USAF	United States Air Force
USCENTCOM	United States Central Command
USEUCOM	United States European Command
USMC	United States Marine Corps
VTOL	Vertical takeoff and landing

Chapter One

INTRODUCTION

OPERATIONS SHORT OF WAR

Operations short of war are one of three functional responsibilities assigned to the U.S. Air Force;[1] the other two responsibilities are the prosecution of war and the mobilization for war. Operations short of war are certainly not new to the U.S. military; they are an integral and honored aspect of U.S. history going back to the nation's beginnings and continuing throughout the Cold War.[2] However, since the end of the Cold War, operations short of war, particularly the subset found in international crises (e.g., Rwanda, Haiti) and lesser conflicts (e.g., Grenada, Panama), have become an increasing part of the U.S. military's operations and concerns. The increase is not just perceived; it is real in the numbers, frequency, and scope of such operations, even though the post–Cold War era is still young.

Ever since the emergence of a "rationalized" defense planning process under Secretary of Defense Robert S. McNamara in the 1960s, at the height of the Cold War, operations short of war have been treated in planning as "lesser included cases": It was then assumed that any force adequate for the major contingencies (global or regional wars) would also be adequate to meet the force requirements for other

---

[1]See *Basic Aerospace Doctrine of the United States Air Force*, United States Government, Vol. 1, March 1992, pp. 259–261, which notes that this particular formulation "of the functions of the Air Force is extracted from Department of Defense Directive 5100.1, *Functions of the Department of Defense and Its Major Components*, which is based on Titles 10 and 14 of the United States Code."

[2]See, for example, Carl H. Builder, "Nontraditional Military Missions," in Charles F. Hermann, ed., *1994 American Defense Annual*, New York: Lexington Books, 1994.

crises and smaller conflicts. Whether or not that approach was ever justified in the past, there are several reasons why it should be critically reassessed now:

- The rapid changes in world commerce, communications, and demographics are changing the character of conflicts and disasters and of U.S. interests, calling for "new" uses of U.S. military forces.[3]

- The lid has come off many simmering conflicts (many of them separatist or ethnic in character) that were ignored or suppressed during the Cold War. Although the United States may not want to become involved in many of these conflicts, few will escape calls for, and the contemplation of, remedial uses of U.S. forces.[4]

- Future U.S. forces will probably be smaller, with less built-in capacity or "slack" for "other" missions, and they are unlikely to retain all of the rich combinations of capabilities that were affordable in the past.

- Some operations short of war will be located where infrastructures are less adequate than those expected for major regional contingencies. At the same time, many of these operations will not warrant either the risks or costs of projecting all of the supporting infrastructures that typically accompany major U.S. force deployments. Under such circumstances, local security, communications, and transportation may pose greater difficulties than those planned (as available or to be projected) for waging war.

All of these developments suggest that operations short of war should no longer be treated as "lesser included cases" for the forces designed to fight the nation's wars. Even though these operations do not now and are unlikely in the future to *size* the combat forces, there

---

[3]They may seem new against the experience of the Cold War, but they are not against the background of more than 200 years of U.S. military history.

[4]Because U.S. society is itself composed of so many different ethnic elements, almost any ethnic conflict elsewhere is likely to reveal U.S. constituencies supportive of one or both sides. The recent intervention in Haiti is an example of a domestic constituency vigorously and successfully pressing for U.S. intervention even though the majority of the U.S. public was apparently not so disposed.

are growing reasons for expecting that a subset of operations short of war could soon size some of the supporting elements and impose some specialized demands upon the organization, training, and equipping of the U.S. armed forces. This report explores those demands, especially for the United States Air Force.

WHAT ARE CALCS?

Operations short of war cover a broad spectrum, from domestic calls for disaster assistance and restoration of civil order to international calls for humanitarian aid and military intervention. Some such operations are relatively short-term, such as relief operations in Bangladesh, relief and evacuation during the Mount Pinatubo disaster, or the evacuations of endangered U.S. citizens. Others are planned as relatively long-term operations, such as the congressionally mandated surveillance of maritime and aerial drug smuggling and the United Nations mandated peacekeeping operations in the Sinai. Still others are of indeterminate length, such as the protection of ethnic minorities in Iraq and Bosnia, awaiting the resolution of political issues.

Collectively, all of these operations short of war have been given a variety of names: nontraditional military missions, military operations other than war (MOOTW or sometimes just OOTW), and noncombat missions—none of which are entirely satisfactory to describe a responsibility of growing size and importance. These operations are not nontraditional in the sense that the U.S. military has been conducting them throughout its history. Nor are they always noncombatant—feeding people sometimes involves keeping them from shooting each other or even from shooting at those who are trying to feed them. Certainly, the air operations to enforce no-fly zones over Bosnia and Iraq have involved shooting. "Operations short of war" is a category so broad as to include such long-term, background operations as military assistance programs to other nations, foreign officer training programs, air transportation for the nation's political leadership, and managing the nation's wetlands, to name just a few of many such activities.

The particular subset of operations short of war that are now stressing the U.S. military forces are not the domestic or routine operations; they are the *international* and *nonroutine* operations short of war, especially those that pose the threat of combat operations. To distinguish this stressing subset from all other operations short of war and to give these nonroutine, international operations a more precise name, we have chosen to call them crises and lesser conflicts (CALCs).[5]

CALCs are defined here as *international* situations involving *non-routine* military operations short of war or preparations for war.

The position of CALCs within the broader span of operations short of war is illustrated in Figure 1. If operations short of war are divided on the basis of whether they are domestic or international in locus, and then further divided by whether they are routine or nonroutine, CALCs are defined by one of the four quadrants. Perhaps the least controversial are the routine domestic operations that have a long

RAND *MR626-1*

	Domestic	International
Routine	Flood control Executive transport Medical support Managing wetlands	Drug interdiction Military assistance Intelligence support Military presence
Nonroutine	Disaster assistance Civil order	Humanitarian aid Peace operations Crisis response Enforcing sanctions Military intervention

▓▓▓ Crises and lesser conflicts (CALCs)

Figure 1—The Span of Operations Short of War

---

[5]For a brief time, the RAND research team called them "lesser regional conflicts" (LRCs); but we quickly learned that there were, in fact, numbered war plans for a few specific lesser regional conflicts (or contingencies); and we elected to avoid any confusion with these specific cases.

history in U.S. civil-military relationships. The most numerous, stressing, and controversial are the nonroutine international operations that make up CALCs. The other two quadrants, like CALCs, are mixed bags; some of the operations included are controversial, as well, but none are particularly stressing on current military forces.

The nature of recent CALCs is shown in Table 1, which lists a decade of nonroutine international operations short of war. A quick inspection of the listings reveals that they are diverse in their locales and purposes. They are numerous and do, indeed, appear to have increased since the end of the Cold War. Furthermore, the list is neither complete nor up to date and does not include the enforcement of air and sea sanctions against Serbia, the air support for United Nations forces on the ground in Bosnia, the peacekeeping forces in Macedonia, the relief efforts for Rwandan refugees, nor the military intervention in Haiti, to name some of the most recent operations.

Just as major regional contingencies (MRCs) have become the dominant conception of war for defense planning purposes, CALCs are becoming the dominant form of operations short of war for defense planning. What is not so clear is whether or not operations short of war, and especially CALCs, will rise in importance relative to MRCs and come to predominate among the peacetime military responsibilities of the U.S. Air Force. Therein lies a potential debate, not just within the Air Force, but also within the entire U.S. military and the body politic.

Lieutenant General A. C. Zinni, USMC, who directed the withdrawal of U.N. forces from Somalia in 1995, addressed this debate in remarks to the participants in a CALC exercise organized by the U.S. Marine Corps:

> The question of whether or not we should be involved in these operations is widely debated. When I sat down and counted up how many I have been involved in during my career, I soon realized that this question has been overtaken by events. We are involved in them; and the question I want to answer is how to do them better.[6]

---

[6]This quote is not verbatim but is the sense of General Zinni's remarks to the participants of EMERALD EXPRESS '95 held at Camp Pendleton, California, April 9–14, 1995.

Table 1

A Decade of CALCs

Dates	Place	Code Name	Nature of Operation
Oct 83	Grenada	Urgent Fury	Noncombatant evacuation
Feb 84	Lebanon	JTF-L	Foreign internal defense
Jun 86	Libya	Eldorado Canyon	Retaliation
1987–1988	Persian Gulf	Earnest Will	Protect sea lines of communication
Jan 88	Haiti	Alpine Bandit	Noncombatant evacuation
Mar 88	Honduras	Golden Pheasant	Border security
Dec 89–Jan 90	Panama	Just Cause	Foreign internal defense
Nov 89	San Salvador	Poplar Tree	Rescue
Nov–Dec 89	Philippines	JTF-Philippines	Foreign internal defense and noncombatant evacuation
May 90–Jan 91	Liberia	Sharp Edge	Noncombatant evacuation
Jan–Feb 91	Israel	Patriot Defender	Missile defense deployment
Apr 91–present	Turkey and N. Iraq	Provide Comfort	Relief for Kurdish refugees
May 91	Bangladesh	Sea Angel	Disaster relief
Jun 91	Philippines	Fiery Vigil	Disaster relief and noncombatant evacuation
Sep 91	Zaire	Quick Lift	Noncombatant evacuation
Sep 91	Haiti	Victor Squared	Noncombatant evacuation
Oct 91	Cuba	GTMO	Relief for Haitian refugees
Feb 91	Former Soviet Union	Provide Hope	Winter relief
May 92	Sierra Leone: Freetown		Military support to U.S. embassy and noncombatant evacuation
Aug 92–Feb 93	Kenya and Somalia	Provide Relief	Relief for refugees
Aug 92–present	Iraq	Southern Watch	Enforcement of a no-fly zone in S. Iraq
Aug 92	Angola	Provide Transition	Foreign internal defense
Aug–Sep 92	Guam	Typhoon Omar	Disaster relief
Nov 92	Bangladesh	Sea Angel II	Disaster relief
Dec 92–May 93	Somalia	Restore Hope	Relief and foreign internal defense
Jan 93	Kwajalein Atoll	Provide Refuge	Relief
Feb 93–present	Bosnia	Provide Promise	Medical support and relief
Sep 93	Haiti	JTF-120	Interdict sea lines of communication
Oct 93	Somalia	JTF-Somalia	Internal security
Jul 94–present	Zaire	Operation Support Hope	Relief

SOURCE: Based in part on Adam B. Siegel and Scott M. Fabbri, *Overview of Selected Joint Task Forces, 1960–1993*, CNA 37 93-0007, Center for Naval Analysis, Alexandria, Va., September 1993, and excerpted in *Joint Force Quarterly*, Winter 1993–1994, pp. 36–37.

THE CALC CHALLENGES

Much of the current public debate about CALCs centers on the circumstances under which U.S. military forces should be employed—particularly a better resolution of U.S. national interests, the mission or purposes assigned to U.S. armed forces, and a definition of limits (in time, objectives, or costs) to their commitments. This debate sharpened in the wake of U.S. involvement in Somalia, where a humanitarian mission to ensure the delivery of food to starving Somalis grew into a military intervention to bring about civil order. The phenomenon has been described as mission (or interest) creep:

> *Interest creep* describes situations in which original national interests in resolving a crisis or conflict—that determine political objectives or the ends sought by U.S. leaders—widen in the absence of conscious decisionmaking. This can happen in coalitions when U.S. objectives fall short of those of our coalition partners or of the United Nations. *Mission creep* is its military counterpart and occurs when the Armed Forces take on broader missions than initially planned.[7]

Whether it was the national interest or the military mission that crept in the Somalia situation can, and undoubtedly will, be argued. But the *missions* assigned to the U.S. military by its civilian leadership are not within the decisionmaking province of the U.S. armed services.

The challenge posed by CALCs for the U.S. military is not so much one of anticipating or determining the nature of future CALC missions as it is of defining *how military power can be used effectively* in a range of difficult situations that are apparent even now (e.g., in Somalia and Bosnia). The Air Force may or may not be able to influence the setting of the nation's objectives or military tasking for CALCs, but it does have an obligation to assess how aerospace power should and should not be used to be effective in a broad range of circumstances that may attend CALCs. Such assessments must, among other things, take into account the following:

---

[7]Anne M. Dixon, "The Whats and Whys of Coalitions," *Joint Force Quarterly*, Winter 1993–1994, p. 28, fn. 1, emphasis in the original.

- "Nobody in the U.S. government has the responsibility, authority, and assets required to plan and execute a prompt, effective response to domestic and international calamities."[8]

- CALCs may often be conducted for very high stakes and under severe political and military constraints.

- It may be more important to *foresee the consequences* of military actions than it is to predict where and why CALCs will occur.

- CALCs may involve politically fragile coalitions and can transmogrify into larger and quite different conflicts.

- Improvements in capabilities for CALCs are as likely (or more likely) to lie in organization, training, and special skills as they are in new equipment or technologies.

- Marginal changes in the forces designed for MRCs are likely to offer significant improvements in the capabilities of those forces for CALCs.

These considerations shaped the research design described below.

RESEARCH DESIGN

The informal objective of the research reported here was "to find the long poles for the USAF in the CALC tent." What were the most important problems the Air Force might face in the future as it responded to CALCs?[9]

Because CALCs are by definition *international* crises and lesser conflicts, it is the regional commanders in chief (CinCs) who must plan and execute the military operations associated with CALCs. From the outset, the research reported here was designed around a series of visits to the staffs at the headquarters of each of the regional

---

[8]John G. Roos, "Help Humanity—Don't Hurt DoD," *Armed Forces Journal International*, September 1994, p. 2.

[9]More formally, in the original project description, the objective of the research was "to enhance the informational basis of USAF initiatives to improve the nation's aerospace capabilities for CALCs by assessing the potential strengths and limitations of aerospace power in CALCs, but with due consideration of other military, political, and economic means."

CinCs. The purpose of those visits was to elicit from the planners and operators responsible for responding to CALCs in their region the current views about the problems and opportunities posed by CALCs. These discussions were deliberately not limited to the problems and opportunities for *aerospace* power in CALCs; however, visits were made to the air component commanders reporting to each of the regional CinCs to ensure that the problems and opportunities for aerospace power were specifically included. The following commands were visited, all during the first quarter of 1994:

- In Virginia:
  — USA Command, Norfolk Naval Base, Va. (January 11, 1994)
  — Center for Low Intensity Conflict, Langley AFB, Va. (January 12, 1994)
  — Air/Land/Sea Applications Center, Langley AFB, Va. (January 12, 1994)
  — Air Combat Command, Langley AFB, Va. (January 13, 1994)
  — Armed Forces Staff College, Norfolk, Va. (January 14, 1994)

- In Florida:
  — Air Force Special Operations Command, Hurlburt Field, Fla. (January 24–25, 1994)
  — U.S. Central Command, MacDill AFB, Fla. (January 26, 1994)
  — U.S. Special Operations Command, MacDill AFB, Fla. (January 27, 1994)

- In Hawaii:
  — U.S. Pacific Command, Camp Smith, Hawaii (February 7–8, 1994)
  — Pacific Air Forces, Hickam AFB, Hawaii (February 9–10, 1994)
  — Special Operations Command Pacific, Camp Smith, Hawaii (February 10, 1994)

- In Illinois:
  — U.S. Transportation Command/Air Mobility Command, Scott AFB, Ill. (February 22–25, 1994)

- In Europe:
  — COMAIRSOUTH, Naples, Italy (March 4, 1994)
  — U.S. European Command, Vaihingen, Germany (March 7–8, 1994)
  — U.S. Air Forces Europe, Ramstein, Germany (March 9–10, 1994)

- In Panama:
  - U.S. Southern Command, Quarry Heights, Panama (March 29–31, 1994)
  - SOUTHAF (FWD), Howard AFB, Panama (March 29–31, 1994)

During these visits, the headquarters' staffs were briefed on the purposes and status of this research and then invited to discuss their views, advice, and concerns about CALCs. The following representative questions illustrate the approach taken by the researchers during the discussions:

1. What kind of crisis or lesser conflict is your command's worst nightmare?

2. What is your greatest concern about your command's preparedness or ability to respond?

3. If your command could have more of any one thing it now has, what would it be?

4. What changes would you like to see in the provision of military (or air power and space) assets to your command?

5. What aspect of military (or aerospace) power would you like to see RAND study so as to improve your command's responses to CALCs?

6. What advice would you like to see us give to the chairman of the JCS (Joint Chiefs of Staff) (or the Air Force chief) about support for your command in CALCs?

Members of the research team[10] were prepared to pursue selected areas—such as aspects of equipment or command and control—in greater depth, whenever possible.

---

[10]The research team visiting the CinCs varied from trip to trip, but one member of the team (Ed Wright) made all of the trips. Typically, five to seven researchers participated on each trip. The research team comprised Carl Builder, Jack Craigie, Steve Hosmer, Dana Johnson, Ted Karasik, and Joe Kechichian of RAND; Colonel Chuck Gagnon, Lieutenant Colonel Mark Anderson, Lieutenant Colonel Ed Wright, and Major George Gagnon of the U.S. Air Force; and Lieutenant Commander Fred Smith of the U.S. Navy.

As the series of visits to the CinCs' headquarters progressed, patterns in the answers were sought and tested, reinforced or rejected. Although there are a few consistent themes across the regional CinCs, each region generally faces a unique set of concerns about CALCs. Moreover, the purpose of the research was not to audit or cross-compare the CinCs, but to ensure that the CinCs concerns were taken into account in the researchers' analyses and assessments of CALC problems and opportunities. Therefore, the observations reported here are, with a few noted exceptions, not CinC-specific and reflect the researchers' appreciation of the national-level problems and opportunities in CALCs. The regional CinC visits did result in the identification of some areas—such as the assignment of military units to the active and reserve components—that needed to be examined more carefully.[11]

RESEARCH DOCUMENTATION

The pertinent literature on CALCs exists primarily in the current press and professional journals. The knowledge and available views on CALCs are growing and evolving rapidly under the relentless pressure of current events. Hence, most of the citations supporting this research are from the contemporary literature. To be sure, CALCs (and operations short of war) have a long and well-documented legacy in military history; but the structural changes in the world, the nation, and the military, just during the past decade, erodes the relevance of much of that history as it pertains to the organizing, training, and equipping of the Air Force or the other branches of the U.S. armed forces. Nevertheless, the recent research, at RAND and elsewhere, on operations short of war was included in this study and has been cited where relevant.

The research by RAND for Project AIR FORCE on CALCs during Fiscal Year 1994 is being documented in two principal reports conceived around two different audiences and purposes:

---

[11]More than 100 specific issues or problems were identified and catalogued by the research team. Many seemed to be of a minor nature or idiosyncratic to the respondent; a few were general complaints which were not in any way restricted to CALCs.

1.   A forthcoming report[12] on the planning and operational prob-
     lems and opportunities posed by CALCs for decisionmakers at all
     levels, from the national command authorities to the service sec-
     retaries and chiefs.

2.   This report on the organizing, training, and equipping problems
     and opportunities posed by CALCs for the military services, es-
     pecially for the Air Force.

Although this report reflects *some* of the views and concerns raised
during the visits to the CinC headquarters, it does not provide a
summary of those visits or our discussions. Such a summary has
been separately prepared by Edgar A. Wright and published as a report for the Air Force.

This report addresses the particular problems and opportunities of
the U.S. Air Force in organizing, training, and equipping aerospace
forces that may be available and provided for CALCs. Of course,
since the Air Force is one of four U.S. armed forces providing forces
for military operations short of war, some of these same problems
and opportunities may apply to other services as well. The authors
have not tried to *limit* their view to Air Force issues; but they have
tried to *include* all aspects that might be pertinent to Air Force planning. This report has been organized around the assigned responsibilities of the Air Force for military operations short of war, which
are described in greater detail below.

In a time of shrinking defense budgets, the financial implications of
changing missions or forces assume greater importance. Although
most of the problems and opportunities explored in this report are
not likely to be "budget busters," any new budget burdens or shifts at
this time can be a cause for someone's concern; and a few of the
ideas associated with equipping the Air Force for CALCs could be
"big ticket" items, with the potential for encroaching on existing development program plans. The authors have not estimated these financial implications at this exploratory stage in the belief that the
current impediments are more conceptual than fiscal. If some of the
arguments contained herein are accepted by the Air Force, then the

---

[12]The author of this prospective report will be Steve Hosmer, a member of the
research team on the project.

authors acknowledge that there will be a need to flesh out those suggestions with more specifics, including resource requirements and implications.

Finally, this report reflects the acknowledged advocacy of the authors for the importance of CALCs to the nation's future security and to the Air Force as one of the nation's military instruments and institutions. It is our hope that this evident advocacy does not vitiate the analyses or devaluate the arguments; but in this time of great changes—in the world and in military budgets—advocacy *for* CALCs may be in shorter supply than are analyses and arguments *against* U.S. involvements in CALCs as potential encroachments upon conventional U.S. military capabilities.

USAF RESPONSIBILITIES FOR CALCS

Under its basic doctrine, "The Air Force is responsible for the preparation of the air forces necessary for the effective prosecution of war and military operations short of war . . . and . . . for the expansion of the peacetime components of the Air Force to meet the needs of war." Preparation of the necessary air forces means to "organize, train, equip, and provide forces" to carry out all the operations required to fulfill the Air Force's primary and collateral functions (now, more commonly called "missions").

Throughout the Cold War, the forces "necessary for the effective prosecution of war" clearly dominated the other two responsibilities for "military operations short of war . . . and . . . for the expansion of the peacetime components of the Air Force to meet the needs of war." The Cold War threats required *ready* forces capable, if necessary, of prosecuting a war to termination in hours or days. Those threats made mobilization and operations short of war secondary considerations in preparing the necessary air forces. So, for more than 40 years, the efforts to "organize, train, equip, and provide forces" has been focused on "the effective prosecution of war," with operations short of war and mobilization being handled as issues on the margins of Air Force priorities. That wartime focus may be contrasted to the peacetime, pre-World War II era when the emphasis was on mobilization and operations short of war.

With the end of the Cold War, the threat has changed dramatically, but the focus of "organize, train, equip, and provide forces [for] the effective prosecution of war" has not. The prospect of war has changed from that of an imminent collision of nuclear superpowers (directly or through their proxies) to that of two possibly simultaneous MRCs, most often illustrated by renewals of the Gulf and Korean wars. In U.S. defense planning, it is now argued that the United States must be prepared for two such MRCs at any one time because an adversary might chose to exploit the U.S. preoccupation with one MRC, as an opportunity to instigate aggression elsewhere. Accordingly, two MRCs have replaced the superpower conflict of the Cold War as the wars the Air Force is preparing itself to prosecute when it organizes, trains, equips, and provides its forces—with operations short of war and mobilization once again being handled as issues on the margins.

Unfortunately, other things beside the threat of war have changed with the end of the Cold War: Defense spending, which had been declining steadily as a portion of federal expenditures throughout the Cold War, now began to decline rapidly in absolute terms as federal deficit spending slowed and post–Cold War priorities shifted toward domestic claims for federal resources. At the same time, operations short of war—particularly in the form of CALCs—seemed to increase, with rising demands to use the U.S. military to solve problems of ethnic conflict, humanitarian and disaster assistance, and civil unrest. The prospects for the rest of this decade are for a continuation of both these trends—fewer resources for the military and more demands for their use in operations short of war—even as the mainstream of U.S. defense planning tries to keep its focus on preparedness for two MRCs.

AN AIR FORCE DESIGNED FOR CALCS

How much different would the forces be if they had, instead, been designed and balanced for CALCs rather than for MRCs? The question is both academic and hypothetical, but that does not mean that the answer is without utility as a planning reference point. To make the question more specific, how would the Air Force *ideally* organize, train, and equip its forces if operations short of war were to become the dominant peacetime responsibility? To answer that hypothetical question is not the same as arguing that CALCs *should* become the dominant responsibility of the Air Force; rather, it is a way of exploring where and to what degree the MRC and CALC demands pull the Air Force in different directions. The question deserves answers if only to calibrate the distance between the two horns of the dilemma that now presents itself. We already know how the Air Force should organize, train, and equip its forces if MRCs were the basis for planning—because that *is* the basis for current planning. But we do not know the answers if CALCs were the basis. It is the *differences* between the two that are interesting, not the extremes as design points.

The authors have exploited that hypothetical question often in this report. As we examine the problems and opportunities for aerospace power in CALCs, we frequently ask and try to answer how and how much different the Air Force might be if it were to organize, train, or equip its forces specifically for CALCs rather than for MRCs. And having done so, as a *gedankenexperiment,*[13] we also back off just as often to explore how the differences between the extremes would have to be realistically split, given the current MRC orientation of the forces.

ORGANIZATION OF THIS REPORT

Chapter Two is a broad examination of the fundamental institutional dilemma that CALCs pose for the U.S. military at this time. In part, it motivates the search for more effective ways of organizing, training, and equipping military forces for CALCs.

Chapters Three through Five separately take up the problems and opportunities for the Air Force in organizing, training, and equipping its aerospace forces for CALCs. Some of the problems and opportunities examined—such as the assignment of units to the active and reserve components—may apply more generally to any of the military services; but others—such as a call for global-range capabilities for insertion and extraction—are offered specifically for Air Force consideration.

---

[13]From the German, literally, a "thinking experiment" or an experiment carried out in the mind as opposed to one conducted in the real world. It was a popular term among European physicists in the first half of the twentieth century to describe how they had arrived at (or how one could arrive at) a particular conclusion.

In Chapter Six, the authors attempt to pull threads from their more detailed explorations and offer some summary observations on what is likely to happen, what the Air Force can make happen, and what actions the Air Force ought to take with respect to its responsibilities toward CALCs as the most demanding subset of military operations short of war.

THE CALC DILEMMA

'We are an army, not a Salvation Army,' Secretary of Defense William Perry recently told Congress. . . . No longer feared as the world's policeman, the U.S. military has become the world's relief worker. Its role model has been turning from George Patton to Florence Nightingale.[1]

The new mission for America's armed forces is still being written, but it is sure to contain essential chapters on humanitarian [and] peacekeeping chores as well as traditional warfighting abilities. Can all that be done in an era of budget cuts? From Haiti to Kuwait, we're doing it right now.[2]

STRESSING OPERATIONS

The U.S. military forces being employed for CALCs are already showing signs of stress. Staffs of forces and headquarters are stretched thin as they try to handle concurrent or successive commitments to CALCs: For the Air Force, the "intense deployment schedules for airlift and tanker crews add more strain on families,"[3] leading to concern for the morale of crews in heavy demand.

---

[1]Ken Adelman, "Dialing 911 for the Military," *Washington Times*, August 12, 1994, p. 19.

[2]Editorial: "Military Isn't Dollar-Short," *USA Today*, November 21, 1994, p. 10.

[3]Steven Watkins and Vago Muradian, "Pushing the Limits," *Air Force Times*, August 29, 1994, pp. 12, 16.

> [Under Secretary of Defense (for policy) designate Walter] Slocombe said he is worried about the burdens placed on "specialized units" that are in heavy demand. . . . "We need to be sensitive to the effect on morale as well as readiness," Slocombe told the [Senate Armed Services] committee in a written response to questions posed prior to his August 10 confirmation hearing.[4]

Of those "specialized units . . . in heavy demand," none has carried a heavier burden than the AWACS.

> AWACS crews [are] one of the Air Force's most overworked and important groups, officials say. Since the end of the Persian Gulf War, AWACS crews have been deployed sometimes more than 200 days per year, according to AWACS crew members.[5]

The Air Force chief of staff described this as an "OPTEMPO problem," where the Air Force has "some folks on the road too much." The solution, he argued, was to limit deployment times to 120 days each year.

> The majority of our people have been deployed less than 120 days over this last year. However, 13 out of 20 of our major aircraft types have exceeded that number. Many of them are special capability aircraft that are in high demand for operations other than war. They include systems like AWACS, AC-130s and rescue HC-130s. These are mission-type aircraft that had their people on the road from 160 to almost 200 days last year.[6]

Units needed for MRCs are committed into operations short of war from which they can not be extracted or recovered quickly. "Soldiers

---

[4] *Inside the Army*, August 15, 1994.

[5] Steven Watkins, "AWACS Crew Unlikely to Face Harsh Discipline," *Air Force Times*, August 29, 1994, p. 4, in an article speculating on the disciplining of AWACS' "airborne controllers involved in the April disaster in which two Air Force fighters shot down two Army helicopters, killing 26 people."

[6] General Ronald R. Fogleman, "Core Competencies—New Missions: The Air Force in Operations Other Than War," presented at the American Defense Preparedness Association Symposium, Washington, D.C., December 15, 1994, as reported in *Air Force Update*, 95-01.

pumping fresh water for Rwandan refugees simply aren't available for rapid reinforcement to Korea, should a crisis erupt there."[7]

Chairman of the Joint Chiefs John Shalikashvili, speaking from his own experiences in the aftermath of the Gulf War, noted that "it has proven difficult for the military to disengage from these operations other than war. For example, what was to have been three to four months of protecting Kurds in northern Iraq has turned into years of U.S. military involvement...."[8]

The loss of training time, devoted mostly to preparing for MRCs, is one concern:

> And after time pumping that water in Goma, our troops may no longer be ready for combat in Korea. Combat readiness dissipates as humanitarian missions rise. Training for food drops and medical distribution differs from training for close air support and tank maneuvers.[9]

> [Secretary of Defense William] Perry pointed out that while U.S. troops are cuddling starving babies and rescuing brutalized adults in far-flung areas, they have scant chance to prepare for their primary mission of fighting and winning the nation's wars.[10]

The effect of CALC operations upon the combat readiness or effectiveness of U.S. forces is not yet certain, but the outlines of the arguments are emerging:

> In a letter to lawmakers the Pentagon made public on Tuesday [15 November 1994, Secretary of Defense] Perry said that the problem was basically too many missions and not enough money. Operations in Rwanda, Haiti, Cuba, and Kuwait cost the Pentagon $1.7 billion more than it expected in the fiscal year that ended

---

[7]Adelman, "Dialing 911 for the Military."

[8]General John Shalikashvili, in an interview with Bob Mahlburg, published as "Shalikashvili Says Aircraft Funding Among Tough Choices," *Fort Worth Star-Telegram*, August 25, 1994, p. 27.

[9]Adelman, "Dialing 911 for the Military."

[10]Ibid.

September 30. To keep frontline units prepared, the Army was forced to raid the budgets of lower-tier forces.[11]

Three Army divisions reportedly had to cancel training to save money. But others see the readiness problem as having deeper roots:

The reality, according to many military officers and analysts, is that the readiness gap reflects a deeper tension between different notions of how best to use the military. At bottom, the readiness issue is . . . a political debate about whether the United States should be performing missions in places like Haiti and Rwanda. If the answer is yes, some . . . say the military should be reconfigured so that it can fight brush fires without causing a serious crunch in resources.[12]

Others say no, the purpose of the military is being distorted:

Notwithstanding all of the changes that have taken place in the world, notwithstanding the new emphasis on peacekeeping, we have this mission: to fight and win this nation's wars. We're warriors. We never want to lose sight of this basic underlying principle. That's why you have armed forces.[13]

Diversion of attention toward CALCs and away from MRCs is also seen as having a potentially adverse effect on performance:

There is little evidence yet that relief missions are degrading combat effectiveness. But some officials point to the recent mistaken downing of two U.S. helicopters over Iraq by two U.S. fighter jets as a warning of the deterioration in performance that can come from extended involvement in peripheral activities.[14]

---

[11]Eric Schmitt, "G.O.P. Military Power Assails Troop Readiness," *New York Times*, November 17, 1994, p. 22.

[12]John F. Harris, "Military Readiness Question Is Founded in Debate Over Roles," *Washington Post*, November 18, 1994, p. 16.

[13]Colin Powell, as quoted by Harry G. Summers, Jr., "First Priorities for the Military," *Washington Times*, November 17, 1994, p. 18.

[14]Bradley Graham, "Pentagon Officials Worry Aid Missions Will Sap Military Strength," *Washington Post*, July 29, 1994, p. 29.

The incident referred to is the April 14, 1994, accident in which two Army Black Hawk helicopters, mistaken for Iraqi Hind helicopters violating the no-fly zone over northern Iraq, were shot down by two Air Force F-15s, killing 15 U.S. citizens, five Kurds, and six military officials from Britain, France, and Turkey.[15] But avoiding peripheral activities and concentrating on training for MRC warfighting is difficult under the current circumstances and policies. "About 48,500 military personnel are currently serving in humanitarian and peacekeeping operations in areas including Iraq, Bosnia, Macedonia, the Adriatic Sea, Rwanda, and the Caribbean Sea for missions involving Cuba and Haiti," according to the Chairman of the Joint Chiefs, General John Shalikashvili.[16]

Some of the airlifts to support CALCs have demanded greater-than-planned use of older, limited-life equipment, such as the C-141s and C-5s. Their remaining life was being husbanded for use in MRCs, pending delivery of their replacements, the C-17. The delays and curtailments in the C-17 program, coupled with the accelerated wear on the older airlifters caused by CALCs, has created a crisis in strategic airlift planning. Undoubtedly, there are other areas in which equipment is being lost or prematurely aged through CALC operations, but the most evident area is in the strategic airlift assets.

CALCs result in the use of forces in operations for which they were not specifically organized, trained, or equipped. Although their proven capabilities make the U.S. military forces the obvious instruments of choice for the U.S. government in responding to international crises, such uses are neither welcomed by, nor without costs to, the military.

> [T]he world has been calling on the Pentagon as its relief force of last resort. The military's gigantic transport capacity and logistical expertise make it well suited to the role, but that does not mean the armed forces always like it.[17]

---

[15]Richard Serrano, "Military on Trial as Pilot Accused in Fatal Downing," *Los Angeles Times* (Washington), November 8, 1994, p. 1.

[16]General John Shalikashvili, in Mahlburg, "Shalikashvili."

[17]Eric Schmitt, "Military's Growing Role in Relief Missions Prompts Concerns," *New York Times*, July 31, 1994, p. 3.

Asked to respond to recent comments by Defense Secretary Perry which could be viewed as indicating the military's less than enthusiastic participation in the African relief mission [for Rwandan refugees, the Vice Chairman of the Joint Chiefs, Admiral William] Owens mirrored his superior's position that fighting men may be somewhat out of place in humanitarian missions. 'We should do everything we can to let leadership know what we can do, but . . . [also what] the impact on our ability to fight will be.'[18]

Clear-cut, fixed military objectives that are both needed and expected for the effective prosecution of wars are often absent in CALCs.

Operations other than war . . . are tricky in that the military lacks a clear compass to guide its use of forces in these missions. While concepts such as using force as a last resort, decisive victory, and application of overwhelming force guide warfighting missions, these concepts have no application to humanitarian missions.[19]

In the same vein as Henry Kissinger's lament about the meaning of strategic superiority during the strategic arms control negotiations, General Shalikashvili asked,

But what does decisive victory in Rwanda mean? I don't know. Do you? It is when we get into these operations, operations we call short of war, that we get uncomfortable.[20]

His discomfort is shared widely enough to have instigated a public debate about the proper role of the U.S. military in CALCs.

THE SHAPE OF THE PUBLIC DEBATE

The debate is generally not about whether there are CALCs or whether they are important; it is about which ones are appropriate for the use of U.S. armed forces. Peacekeeping in Haiti is contentious; protecting Kuwait from a second Iraqi military threat in

---

[18]*Defense Daily*, August 20, 1994, p. 279, second ellipse in the original.

[19]*Defense Daily*, September 2, 1994, p. 354.

[20]General John Shalikashvili, in Mahlburg, "Shalikashvili."

four years is generally not. The difference in attitudes reflects an assumption of incompatibility between missions:

> If the U.S. military becomes perceived as a force that can be enlisted increasingly to do international assistance work while it waits to fight the next war, Pentagon officials fear the strain may lead to diminished combat readiness, mistakes, morale problems, and political trouble.[21]

Some military staff, particularly in the Army, argue that the increasing assignment of the U.S. military to these operations is wrong, that the primary purpose of the military is (or should be) to "fight and win the nation's wars."[22]

> 'We have to strike a balance' between training for war and engaging in other activities, Gen. Gordon R. Sullivan, the Army chief of staff, said in an interview. 'I feel that tension. Everyone has to recognize that the ultimate purpose of the Army is to fight and win the nation's wars.'[23]

General Sullivan's views are echoed by General Shalikashvili:

> The U.S. must not let a growing number of peacekeeping and humanitarian missions distract it from its prime mission of warfighting, Chairman of the Joint Chiefs of Staff Gen. John Shalikashvili said yesterday.
>
> 'My fear is we're becoming mesmerized by operations other than war and we'll take our mind off what we're all about, to fight and win our nation's wars,' he said at a breakfast sponsored by the Association of the U.S. Army.[24]

---

[21]Graham, "Pentagon Officials Worry."

[22]The phrase, "fight and win the nation's wars" is used increasingly to more narrowly define the purposes of the U.S. military. It appears to have its origins in the Army in the post-Vietnam era; but the phrase is now invoked as an enduring truth, obvious to all. However, more than 200 years of U.S. history suggest that the military has always been used to do much more than fight and win the nation's wars. Indeed, it was not until the Mexican War of 1848 that the military was strong enough to fight and win the nation's wars on its own.

[23]*Defense Daily*, September 2, 1994, p. 354.

[24]Ibid.

But not all military people, even in the Army, see it that way. The closer one gets to the people who are responding to CALCs, the less likely one is to find hostility to these operations:

> 'We've seen a real sea change in attitude over the past six months or so,' said a Pentagon official involved in peace operations. 'The services were in denial about the mission; they just wanted it to go away. We've seen a change due largely to the realization that this is where the Army's bread is going to be buttered.'

> Indeed, with U.S. forces more likely to be engaged in places like Somalia, Bosnia, and Rwanda than in another Persian Gulf War, military officials feel nearly as much political pressure to demonstrate relevance as they do readiness. This has led to some interservice jockeying for assignments.

> 'There's a growing realization that if we don't accept these missions, they'll go elsewhere—and so will the forces, meaning the Army will be cut further,' said an Army planning officer.[25]

Even General Sullivan is not immune to this pressure to demonstrate the relevance of the Army to the changing mission spectrum:

> In the recent case of Rwanda, for instance, the Pentagon was about to send one of the Marine Corps' prepositioned squadrons filled with humanitarian supplies to the West African coast when General Gordon R. Sullivan, the Army chief of staff, intervened. According to Pentagon officials, Sullivan made the case that the Army's own prepositioned ships in the region were specially loaded with the kinds of transportation, water purification, and other equipment suited for the Rwanda relief operation. The Army got the job.[26]

It is sometimes difficult to sort out whether the opposition to the use of the U.S. military in some CALCs—particularly those involving humanitarian or peacekeeping operations—reflects concern about U.S. foreign policy or about the proper role of U.S. military forces. For ex-

---

[25]Bradley Graham, "New Twist for U.S. Troops: Peace Maneuvers," *Washington Post*, August 15, 1994, pp. A1, A8.

[26]Ibid.

ample, one editorial[27] begins with the question, "The U.S. military: Are its troops warriors or welfare workers?"—suggesting that the concern is about the proper role for the military:

> Since the end of the Cold War—and especially under the Clinton administration—U.S. armed forces are being turned away from their historic role of defending the nation's security interests and toward a new, thankless and open-ended task of administering global social welfare.

Yet, the next paragraph suggests that the concern is about U.S. foreign policy in its attempt to deal with failed political and economic systems:

> An institution that is trained and equipped to protect this nation and, when necessary, to wage its wars is now being deployed in the world's Somalias and Rwandas to deal with the shambles of failed political and economic systems by dispensing welfare to hapless victims.

Although clearly opposed to this use of the military, the author obviously appreciates that there are good reasons for the U.S. government to turn to its military forces in such situations:

> [They] follow orders and get things done. Thus, whatever the task at hand, whether building tent cities for refugees or dispensing food to starving children, the military has the men, material, discipline, and efficiency to do what failed governments or well-meaning international relief agencies clearly are less capable of doing.

That said, however, it is the *cause* that is seen as not worth the cost:

> Using the military as a social welfare agency is ultimately self-defeating. It places U.S. troops in a succession of situations where no U.S. national security interest is demonstrable. As a result, the U.S. public rebels at the first casualties. Somalia, a humanitarian mission . . . was a textbook case. So long as Americans could watch their troops dispensing food, they went along; as soon as the public

---

[27]Karen Elliott House, "The Wrong Mission," *Wall Street Journal*, September 8, 1994, p. 18.

saw pictures of a U.S. serviceman's body being dragged through the streets of Mogadishu, they wanted their troops home.

And the cost is not just in lives, but in public support for other causes that *are* judged to be worthy of spending U.S. lives:

> Such murky missions will eventually undermine public support for engaging the military even in those situations where national interests are at stake. Worse yet, to cut and run at the first deaths undermines American credibility abroad and encourages the world's aggressors. In the end, the more the U.S. military must dabble in nonmilitary missions around the world, the less likely the public will support its use in another genuine crisis like the Persian Gulf War.[28]

The argument leaves open the question as to which is the problem— undertaking such missions or assigning them to the military. If the latter, who, instead of the military, should or could take on such missions? If the former, what should the U.S. government do about the Somalias and Rwandas in light of significant public pressures from within its own constituencies to act?

Implicit or unchallenged in this debate about CALCs is the assumption that fighting and winning wars is the primary role of the U.S. military. However, the responsibilities of the U.S. military offer no such distinction between war, operations short of war, and mobilization; only the functions (missions) are separated as being primary or collateral for each of the services. In war, there can be no doubt that the most urgent responsibility of the U.S. military is the prosecution of war. The Cold War, by its nature, was a war that had to be prosecuted continuously—through the existence and preparedness of ready military forces. In peacetime, if a threat should loom on the horizon, as in the 1930s, mobilization could once again become a most urgent responsibility of the U.S. military. Today or tomorrow, in a turbulent world of change in many dimensions, operations short of war could become the most urgent of the service's three responsibilities.

---

[28]Ibid.

THE HORNS OF A DILEMMA

CALCs and MRCs are emerging as two horns of a dilemma for the U.S. military. If the future is dominated by MRCs, in planning or in actual operations, the U.S. military will generally have the right kinds of forces, but probably not enough of them. That is not a surprising situation, because the forces emerging from the Cold War were designed for war, but as those forces are reduced through budget contractions, the concerns about their adequacy for war are quantitative more than qualitative. On the other hand, if the future is dominated by CALCs, the U.S. military may have enough resources in the aggregate, but not necessarily the right kinds of forces. Thus, the CALC horn presents problems mostly for the *qualities* designed into the U.S. forces, whereas the MRC horn poses problems mostly for retaining sufficient *quantities*. With declining or even constant budgets, efforts to avoid one horn will only increase the problems associated with the other.

Until recently, the dilemma has been masked. During the Cold War, with larger forces and fewer CALCs, the capacities built into the supporting forces were generally adequate to meet the needs. Defense planning proceeded on the assumption that CALCs could be treated as "lesser included cases" for forces designed to handle one or more wars that might emerge if the Cold War turned hot. Some doubted the validity of that assumption for some kinds of CALCs, even during the Cold War; but the drawdown of the forces—both combat and support—and the increasing numbers of CALCs have laid the horns bare and challenged the assumption that CALCs can continue to be treated as lesser included cases for MRC-designed forces.

The two horns are opposites in more than the scale or intensity of conflicts or in their demands for the quantity and quality of U.S. forces: One is about low-probability, high-consequence contingencies, the other about high-probability, low-consequence situations. One is about the *adequacy* of resources, and the other is about the *allocation* of resources. One could stress the combat forces, the other is already stressing some of the support forces. One raises concerns about the size or adequacy of U.S. forces, the other about their appropriateness or suitability. The conflicting elements create an ugly choice for the U.S. military services: Would they prefer to find themselves in situations for which their forces were

- remotely, but fatally, inadequate, or
- frequently ill-suited to, or inefficient for, their tasking?

Ultimately, the choice must reckon with the relative risks to national interests and to U.S. military institutions, and with the division of the responsibilities for those risks between Congress and the military. Although the military institutions may not be the ultimate arbiters of that choice, they are by no means indifferent to the outcome.

AIR FORCE AT THE CUTTING EDGE

Among the U.S. military services, the Air Force is encountering the dilemma sooner and more severely than the other services because so many of its unique capabilities are in demand and are already stretched thin by concurrent CALCs. These stressed capabilities include the following:

- Airlift, both global and theater, but especially theater, for the delivery of relief supplies and the deployment and support of forces
- Surveillance from the air and space, especially airborne warning and control systems for the enforcement of air security
- Reconnaissance and intelligence from the air and space for situation and risk assessments
- Ground-to-air threat suppression, such as Wild Weasels, for the enforcement of air security.

At the same time, the bulk of the MRC fighting forces—the generic[29] fighters and bombers—are not particularly stressed, for although they are often required for CALCs, they are not required in the *depth* or numbers needed for MRCs. Thus, while some Air Force assets are being stressed by CALCs, others are not, because the forces have been mostly designed and balanced for the sustained combat operations required for MRCs.

---

[29]The term "generic" is used here to distinguish those fighters designed for general purposes, such as ground attack and air superiority, from those configured for specialized capabilities or missions, such as stealth and suppression of enemy air defenses.

Chapter Three

ORGANIZING FOR CALCS

> We're the nation's 911 force, so our concern is not that we're doing these humanitarian missions. We're designed to do those and other missions. Our only concern is [that] when we're called on, we have the resources to accomplish the mission.[1]

> Although Defense Secretary William J. Perry has made a point of appearing responsive to humanitarian and peacekeeping demands on the United States, the bureaucratic structure of his department for dealing with 'operations other than war' remains somewhat schizophrenic.[2]

One of the intellectual devices used throughout this inquiry into CALCs is the hypothetical question: At the limit, what changes would be made in the military forces if CALCs rather than MRCs were the dominant design consideration? Exploring the answer is more useful in suggesting directions for potential change than it is in arriving at any realistic or expected end point. Indeed, even at the end point, this research has uncovered nothing to suggest that CALC capabilities would benefit from any changes in the basic Air Force organization by commands, wings, and squadrons. However, changes might be indicated (a) in the location of those squadrons or wings between the active or reserve components, for achieving readiness; (b) in geography, between the regional CinCs and the Air Force commands;

---

[1]Schmitt, quoting Major General Thomas L. Wilkerson, a senior Marine Corps planner, in "Military's Growing Role."

[2]Graham, "New Twist for U.S. Troops."

and (c) in a headquarters point of advocacy, for the enhancement of CALC capabilities.

Organizing for CALCs is, of course, more than an Air Force issue; it is an U.S. national and international military issue, even though it is the Air Force perspectives that we mostly pursue in this report. In urging consideration of organizational changes to enhance Air Force capabilities for CALCs, we are mindful that CALC problems need to be resolved, on a broader scale, throughout the U.S. military and within the international community. The Air Force, however, could be the example or the institutional leader for larger organizational initiatives.[3] With that larger challenge in mind, the following organizational issues are but a starting point.

ACTIVE/RESERVE ISSUES

> [W]e may need to redistribute some force structure between the Guard or Reserve and active-duty units. The example that comes to mind is our rescue assets. We have a significant portion of this force structure in the Guard and Reserve. They deploy to support our contingency [operations] as often as practical. But there is a limit on the amount of volunteerism we can expect from our citizen-airmen. The result is that our active-duty stateside HC-130 crews were deployed 194 days last year. That's far too much.[4]

High on the list of organizational issues raised by CALCs is the question of the appropriate balance between the active and reserve components. It is a question less about the relative size of than the kinds of forces that should reside in each component. Should the combat forces reside mostly in the active component, with the support forces mostly in the Reserve and Guard?

> Among the [restructuring] questions to be answered: Will the reserve components, especially the Guard, continue their shift away from glamorous fighters into workhorse tankers and airlifters?

---

[3]The authors are indebted to RAND colleague Milt Weiner for the suggestion that we consider the Air Force and its organizational issues as setting the stage for a larger cast of national and international actors that must be involved in improving CALC capabilities.

[4]Fogleman, "Core Competencies."

'The model that may be appropriate for the future—a continuing series of ongoing crises—that's a fundamentally different model than we've designed our reserve forces to accommodate,' said RAND researcher Bruce Don. 'The current move, which moves fighters out of the Reserve into the active, may be out of consonance with this.'[5]

If CALCs were the principal basis for designing the forces, the biggest organizational change would probably be in the assignment of the three functional responsibilities between the active and Reserve/ Guard forces: The prosecution of war would once again, as it has before in times of peace, become the principal responsibility of the Reserve/Guard units, under the mobilization leadership of the active component. And operations short of war would once again dominate the routine operations of the active forces. Obviously that kind of radical change—at least from the Cold War perspective—is not likely to occur without a major shift in national security think- ing.[6] The pivotal point of such a shift is the selection of big wars or of little wars as the principal and urgent diet of the U.S. military.

However the services ultimately choose to tailor their forces, it is clear that the shift in emphasis from a large-war scenario to one in- volving smaller, but more frequent, operations other than war is forcing reconsideration of issues fundamental to the role the re- serves will play in the future.

[T]he armed services are wrestling with the . . . issue of how to shape their forces to deal with an international landscape that is increas- ingly punctuated by regional and ethnic conflicts. The peacekeep- ing operations and humanitarian assistance missions necessitated by these types of conflicts depend heavily on the U.S. military's lo- gistics infrastructure, and the majority of these capabilities reside in the Reserves. This is especially the case in the Army.

The Army's dependency upon the Reserves for its supporting capa- bilities has been deliberate and explicit. By locating more of the sup- porting forces in the Reserve/Guard component, the Army is able to

---

[5]Andrew Compart, "'Decade of Change," *Air Force Times*, October 18, 1993, pp. 20, 24.

[6]The radical nature of this change is explored in Carl H. Builder, "Looking in All the Wrong Places?" *Armed Forces Journal International*, May 1995, pp. 38–39.

keep the active "tooth-to-tail" ratio high, which keeps within the favored active force more of the favored combat arms (armor, infantry, and artillery) and, as conceived in the aftermath of the Vietnam War, ensures that the nation is mobilized (i.e., supporting its Army) before committing its military forces to war again.

Although the Army has probably been more determined than any of the other services in pursuing that shift, it also has by far the largest ratio of reserves to active forces of the four services.[7] Nevertheless, the Air Force ratio of Reserves to active forces—which is only one-third that of the Army's—also reflects a shifting of the supporting forces to the Reserve/Guard:

> In crews and aircraft combined [the Guard and Reserve] are responsible for more than half of the Air Force's airlift capability. The Reserve component also is responsible for a large percentage of the air refueling mission, which supports airlifters. . . .[8]

The trends in the Air Force that have led to that balance are shown in Table 2.[9]

Table 2

Trends in Reserve/Guard Burdens

	Air Force Guard and Reserve Contributions (percent of total force)[a]		
	1987	1990	1993
Tactical airlift	n/a	64	64
Strategic airlift	11	17	22
Air rescue	36	75	71
Air refueling	21	24	41
Weather reconnaissance (Reserve)	28	40	100
Tactical reconnaissance (Guard)	54	60	100

[a]Shown by year is the percentage of the Air Force's total force supplied by Air National Guard and Air Force Reserve members.

---

[7]"Can the United States Increase Reliance on the Reserves?" *RAND Research Brief,* Santa Monica, Calif.: RAND, RB-7501, September 1994.

[8]Andrew Compart, "Job Can't Be Done Without Reserves," *Air Force Times,* August 15, 1994, p. 13.

[9]The table is adapted from Rick Maze, "Units Wield Political Clout," *Air Force Times,* October 18, 1993, p. 28.

Particularly dramatic is the extent of the shifts to the Reserve/Guard of the capabilities frequently in demand for CALCs—rescue and reconnaissance—and the balances that have been struck for tactical airlift and air refueling.

It is evident that the armed services are experiencing problems with the current division of responsibilities between the active and reserve forces For the Haiti intervention, it was necessary to turn to the Reserves for substantial numbers of volunteers from supporting units:

> Mr. Perry said the [1,600] reserve forces called up by the Pentagon [for duty in Haiti] will be used to fill out active duty units but do not include general-purpose combat forces. . . . Reservists' military specialties include tactical airlift, aerial port operations, military police, medical support, and civil affairs. . . . Civil affairs units are specialized in such matters as languages, law, and the operation of electrical, water, and sanitation systems.[10]

Even though the services are finding that the supporting units they need most for CALCs are often located in the Reserve/Guard—where they may not be immediately ready or accessible—the corrective options are more easily described than acted upon:

- Get more of the supporting units out of the Reserve/Guard and into the active forces; or

- Make it easier to gain access to (call up or employ) the supporting units in the Reserve/Guard.

The difficulties are both institutional and political. For example, the first

> . . . option is to shift more of the services' logistic support capability into the active-duty force. This idea [applies] particularly well in the Army, which has about 70 percent of logistics capability in the Reserves. . . . But Army warfighters have traditionally vetoed the idea of trading combat structure for logistics structure—especially as the Army's overall size has been reduced to ten divisions. Many in the

---

[10]Bill Gertz, "Clinton OKs Call-up of Reserves," *Washington Times*, September 16, 1994, p. 16.

> Pentagon believe that the Army is already so short on 'tooth' that it cannot afford to increase the 'tail' end of its 'tooth-to-tail' ratio.[11]

Many in the Air Force would probably make the same argument: As the forces are drawn down, it is important that the "tail" not be allowed to increase at the expense of the "teeth."

The second option is to make

> ... greater use of the reserves in the future. But the political intricacies inherent in calling up the reserves make it unlikely that any administration is going to make frequent use of the presidential selective reserve call-up—particularly for peacekeeping and humanitarian assistance operations that may not be perceived as central to U.S. national interests.
>
> One way to improve the Pentagon's ability to make use of the reserves and to avoid some of the politics of a full reserve call-up is a plan the Pentagon has proposed that calls for limited authority by the Secretary of Defense to call up 25,000 reserves. The Pentagon submitted the proposal to Congress this year, but it has not fared well to date. Proponents of this plan contend that it would give the services greater access to specific types of reserve forces that are needed for the types of humanitarian assistance missions and peace operations that U.S. armed forces are expected to undertake in the foreseeable future.[12]

A more recent proposal for gaining access to the Reserve/Guard for CALCs is to assign these units to CALC duties during their routine annual two-week training period. That proposal, while superficially appealing, leans on what the Reserve and Guard have the least to offer—their limited time—in an effort to relieve active units with what they need most—the supporting skills now found in the Reserve/Guard. Even if Reserve/Guard units could devote all of their annual two-week training period to CALC duties (i.e., no travel or work-up times), it would take 26 Reserve/Guard units to fill in for one full-time active unit. Rotation of the active units, to reduce the length of their overseas deployments, would reduce this ratio con-

---

[11]Margo MacFarland, "Rethinking 'Tooth-To-Tail,'" *Armed Forces Journal International*, September 1994, p. 45.

[12]Ibid.

siderably; but, even with deployment times limited to 90 days each year for active units, the required ratio of Reserve/Guard to active units would still exceed six-to-one. And that simple arithmetic does not account for the increased airlift burdens of moving units every two weeks.

The Air Force has been able to gain substantial access to its Reserve/Guard units through volunteerism, but

> No one has studied in depth the limits of reserve participation— how much time off civilian employers of part-time reservists are willing to give their employees each year, or how much separation their families can take. . . . Guard and Reserve leaders believe their members are moving close to the breaking point and cannot be asked to serve more days.[13]

This consideration suggests that some of the supporting units so much in demand for CALCs should be moved into the active component and, if necessary, some of the depth in the combat forces—particularly the unspecialized day fighters and bombers—should be moved into the Reserve/Guard units. That suggestion should not be taken to the extreme—CALCs do require combat forces, but not in the depth of numbers currently maintained to wage a wartime campaign against the enemy. The combat forces required for CALCs are more likely to be the specialized units, such as those for the suppression of enemy air defenses (SEAD), the delivery of precision guided munitions (PGMs), and stealth.

One cost of such a shift of combat and support elements between the active and reserve components would be in readiness for war, not in dollars. Indeed, the shift might result in dollar savings because of the reduction in the tempo of combat training in exchange for increased support operations and training. Although any reduction in combat readiness for war would be seen as highly undesirable to many or most of today's military leaders, it must be remembered that the premise underlying that suggestion is quite different from the situation that those leaders see today. So long as MRCs and the retention of combat forces dominate military planning, the shift suggested here between the active and Reserve/Guard components seems most

---

[13]Compart, "Decade of Change."

unlikely. Such a change would probably come about only if there were a radical shift in national security priorities—something that seems more likely to be imposed from the outside by a new administration or by Congress rather than implemented from within the U.S. defense establishment.

Nevertheless, the Air Force may be in a better position than the Army if such a shift is imposed by circumstances or directive. The Air Force's relationships between the active and Reserve/Guard components appear to be significantly different from those of the Army. When one of the authors asked a senior Army officer for an explanation of the seemingly large differences between the Army and the Air Force in their relationships between active and Reserve/Guard components, the following was offered:

> The active Army's problem with the Reserve/Guard units is not with the soldiers, but with their leadership. The experience levels of the active and Reserve/Guard soldiers is not all that different. Indeed, many Reserve/Guard soldiers have had prior experience in active units. But the leaderships, after 15 to 30 years in service, have vastly different experiences. There is no way that a part-time senior officer in the Reserve/Guard can gain the . . . full-time experience of an active duty officer who has had many different assignments throughout his career. It is the Reserve/Guard leadership that the active elements do not trust. If the active force could remove the Reserve/Guard leadership upon mobilization, there would not be a problem. Indeed, that is why the active elements would rather absorb Reserve/Guard units into their ranks in smaller units (e.g., as battalions) rather than see them mobilized as divisions or corps.

> The Air Force's problem is different. The Army fights as divisions, with very large numbers of soldiers depending for their lives upon the competency of its senior leadership. The Air Force typically fights in much smaller units, with fewer lives on the line, and those few are much more dependent for survival upon their own competency than that of their senior leadership.[14]

---

[14]A paraphrasing of a luncheon conversation between the senior author and an active duty Army officer at Carlisle Barracks on October 31, 1994. For obvious reasons, the officer's identity is withheld.

If that assessment is near the mark, then the Air Force's superior relationship between its active and Reserve/Guard components is not just a fortuitous accident of institutional history, but a consequence of the medium (air) and technical means with which it fights. Thus, if any of the U.S. armed forces has the opportunity to make uncomfortable shifts between its active and Reserve/Guard units to improve its CALC capabilities, it is the Air Force.

The quality and commitment of the Air Force Reserves is apparent to the CinCs, who are the end users of the forces: General George Joulwan, referring to the potential CALC overload of his forces in the European Command, observed, "We're getting very, very good support from the reserve forces, particularly the Air Force. This helps offset some of the forward-based force we don't have."[15]

ACTIVE FORCE ISSUES

For CALCs, the active units need to be configured for deployments in smaller slices to meet the needs of a greater number of geographically dispersed operations, as opposed to being concentrated for one or two MRCs. Although centralized organizations make sense for MRCs, CALC capabilities are likely to be enhanced by decentralized organizations. In that respect, as in others, MRCs and CALCs present directly opposed incentives and imperatives.

First, the regional CinCs need at least some of the centralized assets, such as theater transport and reconnaissance capabilities, stationed forward in their theaters so that they can "lean forward" with lead elements when crises emerge. Several of the CinC staffs interviewed for this research offered the opinion that the centralized management of some "national" assets, such as space, reconnaissance, intelligence, and strategic airlift, has hindered rather than helped in the timely responses of their CinCs to CALCs. This is not to imply that each CinC should have a full "kit bag" of resources to deal with all crises; but it does imply that centralization of theater transport and reconnaissance can immobilize and blind the CinCs at a time when they could be "bunching their muscles" to respond in a timely fash-

---

[15]Tony Capaccio, "Supreme Allied Commander Sketches Challenges of 'New NATO,'" *Defense Week*, Vol. 16, No. 6, February 6, 1995, p. 8.

ion. While that need of the CinCs is not an Air Force but a joint issue for resolution, the Air Force is a supplier of the capabilities in greatest demand; and its posture will bear upon the options advanced and adopted for all the services.

More generally, one of the organizational problems of a force designed primarily for MRCs is units that may be too large to operate efficiently in CALCs or to serve the standby needs of the CinCs. An Air Force designed to deploy and fight as a wing or squadron may provide more than is needed in many CALCs. If there is a need to split a squadron's primary equipment between two or more geographically separated CALCs, can the maintenance and other supporting functions also be divided? An Army officer put it this way:

> The Army fights by divisions. When the Army thinks of its logistical needs, it thinks in terms of division slices. The Army knows how to supply a division with artillery shells, but we do not know what a battalion slice looks like. There is no such thing.[16]

Thus, one of the organizational issues that CALCs pose is whether some Air Force units need to be designed for more fragmented operations in CALCs. Some obvious candidates are units of airlifters, AWACS, reconnaissance, and SEAD that may be advantageously deployed forward to CinCs or CALCs in flights of two to four aircraft. Configuring command and supporting elements for such fragmented deployments is probably the main challenge. Not all squadrons, of course, need to be so configured; but those that do will probably require larger numbers of people in their command, staff, and supporting elements (e.g., more tail for the same teeth).

A related question is whether the principal equipment deployed forward in small elements needs to be rotated with its crews or can be left in place, with only the crews being rotated. The AWACS experience since Operation Desert Storm should be pertinent to the answer. When aircraft were cheap and plentiful, rotating planes and crews probably made sense; but in the new environment in which aircraft are expensive and few and the geographical demands for their use are diverse, past practices need to be examined.

---

[16]Paraphrased from a discussion of CALC problems with the U.S. European Command staff in Vaihingen, Germany on March 7–8, 1994.

Similarly, crew ratios for high-demand assets probably need to be re-examined. Such an examination for AWACS crews has probably already been precipitated by the accidental shoot-down of two helicopters over northern Iraq in April of 1994. But the stressed systems do not stop with AWACS. There is evidence of crew stresses for air-lifter and tanker units as well: "The intense deployment schedules for airlift and tanker crews add more strain on families."[17] Until SEAD capabilities are stabilized in active units, the Reserve/Guard units so equipped are also being stressed. Whether this is a crew ratio problem or another imbalance between the active and Reserve/Guard for CALCs is an issue worthy of more careful examination.

> The 124th [Fighter Group] at Boise Air Terminal is . . . testing the limits of Guard and Reserve participation in the total Air Force . . . because it is one of only three units in the entire Air Force flying the F-4G 'Advanced Wild Weasel'—and soon will be one of only two. The F-4G is the only aircraft dedicated to searching for and destroy-ing enemy air defenses, such as surface-to-air missiles, and it has been in heavy demand since it proved its value during the Persian Gulf War.[18]

> The Guard's 124th Fighter Group out of Boise Air Terminal in Idaho is on its second six-month deployment with F-4G 'Advanced Wild Weasels' in Saudi Arabia.[19]

AIR CONSTABULARY CAPABILITIES?

A more pointed organizational question posed by CALCs is whether or not active Air Force units should be specialized for CALCs. Although that question was seriously considered in this research, an affirmative answer was not obvious, even in a hypothetical world dominated by CALCs rather than by MRCs. The only area in which the possibility of specialized CALC capabilities seemed worthy of

---

[17]Watkins and Muradian, "Pushing the Limits."

[18]Andrew Compart, "When Part Time Stops Being Part Time," *Air Force Times*, May 30, 1994, p. 20.

[19]Andrew Compart, "How Capable Are Fighter Units?" *Air Force Times*, October 18, 1993, pp. 20, 24.

consideration was in the provision of ground security from the air, such as that now being provided by the Air Force in Operations Deny Flight (Bosnia), Provide Comfort (Turkey and Northern Iraq), and Southern Watch (Iraq). These operations, designed to protect those on the ground from the air only if and when they are threatened, have the character of air *constabulary* functions. They differ from normal air combat operations in that they typically employ air power in combat only in response to hostile acts and then only under restrictive rules of engagement. Those conditions are foreign to the traditional uses of air power that have been more typically employed in planned campaigns in which the initiative to strike resides with the airmen.

The use of air power as a constabulary is not without precedent. In the 1920s, Hugh Trenchard, the father of the RAF, saw air constabulary capabilities as a way of demonstrating the effectiveness of air power to the British public and government.

> As part of the settlement of World War I, Britain had accepted from the new League of Nations a supervisory 'Mandate' for a clutch of new 'nations' formed from the territory that had belonged to the Turks. These included Palestine, Transjordan, Mesopotamia, the Lebanon, the Hejaz, and the Yemen, all of which were squabbling with themselves and the outside world as they still do today. In 1920, for example, quelling rebellion in Mesopotamia cost the British 2,000 military casualties and £1,000,000. Trenchard proceeded to demonstrate that the Royal Air Force, even though shrunk from 96 squadrons in France at war's end to only 25 1/2, could handle Britain's problems in the Middle East effectively and at far less cost. He then did the same thing on the troubled Northwest Frontier of India. By 1924 . . . efforts to disband the RAF had disappeared, and Trenchard was secure in the reputation he carried ever after as its 'Founder.'[20]

The issues raised by air constabulary capabilities are more likely to be found in training (e.g., reacting to violations) and equipment (e.g.,

---

[20]James Parton, "The Thirty-One Year Gestation of the Independent USAF," in *Aerospace Historian*, Fall, September 1987, pp. 151–152. Bruce Hoffman, in *British Air Power in Peripheral Conflict, 1919–1976*, Santa Monica, Calif.: RAND, R-3749-AF, October 1989, pp. 4–35, provides an excellent description of these colonial operations of the RAF.

long-term surveillance) than in organization. However, if CALCs should ever warrant their own specialized organizations (something the authors doubt), the first candidates might be units configured, trained, and equipped specifically for air constabulary functions and capabilities—providing security for those on the ground through the uses of air power.

WORKING WITH OTHERS

An Air Force organized for CALCs would probably forge more intimate and sustained ties with other organizations—not only with relief and humanitarian nongovernmental organizations (NGOs), such as the International Red Cross and Doctors Without Borders, but also with pertinent governmental organizations, such as the World Health Organization and the International Atomic Energy Association, whose specialized knowledge or skills may be essential for CALCs. The need for such ties is not just at the highest political or military levels; it extends down to the armed services and the tactical units they deploy into CALCs:

> In the new world disorder, many operational situations facing U.S. and allied forces have become increasingly complicated by domestic, economic, and environmental—as well as military—considerations. Unified actions in these situations require military forces to coordinate efforts at the operational and tactical levels with both governmental and nongovernmental agencies. In many instances, relationships among joint and combined task forces and these agencies will be ill-defined until liaison is effected. Moreover, relationships are likely to vary with each agency. Nevertheless, involvement by governmental and nongovernmental agencies, in coordination with military action, is likely to be integral to crisis resolution.[21]

The development and sustainment of such ties would be an appropriate task for the Air Force office proposed at the very end of this

---

[21]Thomas C. Linn, "The Cutting Edge of Unified Actions," *Joint Force Quarterly*, Winter 1993–1994, pp. 34–39, quote from p. 38.

chapter as a point of advocacy for improved CALC capabilities.[22] Many of the ties might be not much more than a contact point—a name and number—in another organization, to which the Air Force could turn quickly for information or assistance. Of course, some of these ties would be most effectively forged through other U.S. governmental agencies, such as the State Department and the Department of Energy.

The particular agencies that can usefully contribute to CALCs and their relationship to the military units involved will vary widely with the circumstances. The storm relief efforts in Bangladesh are illustrative:

> The kind of the crisis at hand will determine the nature of the involvement of the agencies. In Sea Angel, which provided disaster relief in the aftermath of a cyclone in Bangladesh, the JTF coordinated its efforts with the Department of State and the Agency for International Development with which memoranda of agreement existed. It also developed ad hoc relationships and a division of labor among the International Red Cross, Red Crescent, CARE, Save the Children, and other relief agencies. While many nongovernmental humanitarian organizations eschew the appearance of formal relationships with military forces, they have nevertheless become dependent on them for security and even logistical support.[23]

A good example of the growing dependency of relief organizations for military logistical support are the efforts to save the Rwandan refugees who fled to Zaire:

> But disease and death [in Goma, Zaire], for all their seeming relentlessness, were finally slowed in large part by the same sorts of military skills that distinguished the allied victory in the Persian Gulf War: the ability of military personnel to organize the supply of wa-

---

[22]The Army's efforts to develop closer working relationships with NGOs are described by Andrew Weinschenk, "In Rwanda's Wake Pentagon Draws Closer to U.N. Relief Groups," *Defense Week*, Vol. 15, No. 47, November 28, 1994, pp. 1, 12.

[23]Linn, "The Cutting Edge of Unified Actions."

ter, shelter, sanitation, medicine, and food for vast numbers of people over great distances under challenging conditions.[24]

Some observers of the Rwandan disaster went further and offered explicit comparisons between the combat and support capabilities that the military brought to bear in that crisis:

> What the world's relief agencies need, they now believe, is more logistical support from the world's armies. Relief experts have in mind not so much the high-profile, and risky, French military presence in Rwanda. . . . Rather, the experts are more encouraged by the success of the discreet noncombat support provided here by the Americans, Irish, Israelis, and Dutch.[25]

> 'The military's logistics are best suited for addressing the problems we face in these humanitarian crises,' said Joelle Tanguy, the executive director of the American office of Doctors Without Borders, an international relief organization. 'Providing airport handlers, tanker trucks, pumping stations—they can do it much better than agencies who don't have the tools or the programmatic experience.'[26]

Of course, such praise for military logistical support is seen as a two-edged sword. It demonstrates the competency and worth of the military services in CALCs, but it also raises anew questions about who should provide such logistical capabilities—the military or the relief agencies?

> America's war planners fear the results of their success. They are concerned that as they improve their skills in organizing large-scale operations (as they did in Somalia), devising ways to parachute supplies in treacherous areas (as in Bosnia), and speeding supplies to disaster sites from the United States, Europe, and the Pacific, the world will increasingly turn to Washington rather than develop a dedicated international force to provide relief.[27]

---

[24]Jane Perlez, "Aid Agencies Hope to Enlist Military Allies in the Future," *New York Times*, August 21, 1994, p. IV–6.

[25]Ibid.

[26]Schmitt, "Military's Growing Role."

[27]Ibid.

Although there may be some merit in creating a dedicated international relief force to take up these duties, the military services can hardly wait for the event. Indeed, it seems unlikely that such a separate relief force could be made robust enough to meet the needs that arise in extreme events—just as domestic police can generally not be made robust enough to withstand widespread urban disorders. So, even if an international relief force is created, the U.S. military must remain prepared to respond to CALCs, including relief operations, as ordered; and, when ordered, to do their very best. The relief demands on the military might be fewer; but they are not likely to go away entirely unless the scope of human disasters can somehow be limited. The trend in disasters, unfortunately, is exactly the opposite.

Nevertheless, an international relief force would reduce the demands on the U.S. military for many of the smaller disasters; and that should be in the interest of the Air Force to foster in parallel with its own preparations. The U.S. military could "export" much of its logistical know-how and, perhaps, even surplus equipment to an international relief force with the beneficial effect of reducing the "wear and tear" on its own people and equipment for small operations scattered around the globe.

A POINT OF ADVOCACY

Compared with changes in training or equipment, organizational changes could offer the Air Force some of the least-cost, highest-payoff responses to the problems posed by CALCs. But organizational changes are likely to be among the most disturbing to the Air Force as an institution and, therefore, could be the most difficult to effect. For these reasons, one of the most important changes in organization for the Air Force to consider is the creation of a headquarters point of advocacy for CALC capabilities.

Currently, there are no eyes, ears, or voice within the Air Force to watch, to listen, or to speak for opportunities to improve USAF CALC capabilities—even where those improvements might be achieved at modest cost or with modest changes on the margins—not just organizational changes, but also in training and equipment. The current senior leadership of the Air Force, forged as Cold Warriors and drawn mostly from the combat rather than the supporting elements, must overcome considerable experiential and interest barriers to give

CALC capabilities much attention. And, in the absence of much attention, CALC capabilities are likely to be overlooked in the presence of urgent efforts to retain the mainstream combat forces against budget pressures.

A point of advocacy for CALCs might take the form of a four-letter office (division) in the Air Force headquarters, under the Directorate of Plans, Deputy Chief of Staff Plans and Operations (AF/XOX). That office[28] could serve as a lightning rod throughout the Air Force for initiatives to improve CALC capabilities. The concepts offered in this report might serve as a "starter set" of ideas to explore with appropriate commands and agencies throughout the Air Force and, as appropriate, with the regional CinCs, Joint Staff, the DoD, and other governmental and nongovernmental agencies.

Without a point of advocacy to take the lead in exploring, evaluating, and promoting opportunities to improve USAF CALC capabilities, few of the other suggestions or proposals offered in this report could be expected to find fertile ground or a chance to grow. Thus, the first and most important recommendation made in this report is an organizational one: Create a point of advocacy for USAF CALC capabilities within the Air Staff. All of the remaining recommendations or suggestions made here—whether for organization, training, or equipment for improved CALC capabilities—will depend upon the existence of such an advocacy point if they are to become anything more than passing ideas.

---

[28]The authors have not explored the cost of such an office, in the belief that the Air Force itself is the best judge of where such an office should be located in the USAF organization, how large it should be, and what other capabilities might have to be sacrificed to create the office.

Chapter Four

TRAINING FOR CALCS

> Shalikashvili rejects the notion . . . that units should be trained exclusively for peacekeeping. When what starts out to be peacekeeping turns into violence, trained warfighters are what is needed. Therefore, the best peacekeeper is the best soldier, sailor, or airman. . . .[1]

> The military's prime focus should continue to be on training to meet the national military strategy of fighting and winning two wars occurring nearly simultaneously, the JCS chairman said. Neglecting training in warfighting for peacekeeping missions would be an 'awful mistake,' Shalikashvili said. The challenge for the military is to determine how to add training in peacekeeping without affecting training in warfighting.[2]

Although peacekeeping is not the essence of CALCs, it is clearly one of the many military skills—along with logistics, medicine, transport, security, and so forth—that are frequently required in CALCs. And the challenge posed by the JCS chairman not only remains standing, it can properly be broadened when applied to CALCs: Is the "challenge for the military to determine how to add training" for CALCs "without affecting training in warfighting"? The issues nested within that broader challenge are several:

---

[1]General John Shalikashvili, in Mahlburg, "Shalikashvili."

[2]Ibid.

- Do CALCs require *additional* training, or do they require the adaptation of existing military skills? Is being the "best soldier, sailor or airman" the best way for the military to serve in CALCs?

- Do CALCs require *any* training, or do they require only education?

- Should warfighting skills take precedence over *all* other military skills?

- Would training for CALCs *necessarily* detract from MRC training?

As might be expected, these issues are more easily raised than resolved. When posed as direct questions, the answers seem to be both yes and no:

- CALCs may require some additional training, but they are mostly an adaptation of existing military skills. Being the best airman is a good start, but it may not be enough for the Air Force to be most effective in CALCs.

- CALCs may not require much in the way of additional training, but they may require somewhat more in the way of additional education.

- Proficiency in fighting skills is maintained through practice, and in the absence of war, training in warfighting skills is the substitute for practice. Hence, warfighting training must take precedence over CALC training. But CALCs may not require as much in the way of training as they do in education; and education tends to be a one-time demand, not a steady drain, on practice time.

- Proficiency in flying skills must also be maintained through practice, and in the absence of war, CALC operations do provide flying practice—albeit different from that of war or warfighting training and depending on divergences between wartime and CALC uses of different aircraft.

Thus, the impact of CALCs upon MRC training and the cross-utility of CALC and MRC skills is, at the least, a mixed bag that deserves more qualified judgments than absolute dictums. The good news is that training may be the least critical of CALC problems. For the USAF leadership, it is clear that the training issue raised by CALCs is not so much what training they require as it is how CALCs are affecting the training needed for MRCs. Air Force Chief of Staff General Ronald R. Fogleman put it this way:

> The high operations tempo of deployments and overseas operations for humanitarian, disaster relief, and peacekeeping efforts may eventually degrade military readiness and training. . . . Contingency operations over the past year in Bosnia, Rwanda, Cuba, and Haiti have had nominal affect on airlift, tanker, and sealift crews, but are taking a toll on airdrop aircrews. . . . As our aircrews get swept into something like the African relief effort, they are deployed away from home for a long period of time and they do not get the opportunity to stay current in their airdrop capabilities. . . . When the crews return home, we can't give them time off, because airlift capability is so critical. [Aircrews are immediately entered into] an extensive training program to get them back up on the step.[3]

The military participants in CALCs are not likely to complain that they are inadequately trained for the tasks they face; they are more likely to complain that CALC duties interfere with their "normal" training. Whether either of those postures is strictly correct is less obvious: Some training specifically for CALCs almost certainly would not hurt; and it may be difficult to demonstrate that CALC duties have really detracted from the skills that people will end up needing in a very uncertain future.

A more insidious problem has surfaced, specifically with AWACS, which are in constant and heavy demand for CALCs. Secretary of the Air Force Sheila E. Widnall addressed the training issue for AWACS this way:

> We're starting to pay more attention to this whole issue of the relationship between flying operational missions and staying current, which are not the same thing. . . . We saw that in the AWACS incident [the accidental downing of two Army helicopters]. There's the question of keeping comprehensive training going while you're in-

---

[3]Air Force News Service, October 12, 1994.

volved in operations that exercise only a small fraction of your talent.[4]

However much CALCs may interfere with the "normal" MRC training for AWACS crews, they are now interfering with the training of new and additional crews to relieve the stress on the CALC-deployed crews. The demand on AWACS is so great that there are not enough aircraft left to meet the higher training demands required to increase the crew ratios.[5] Clearly, AWACS now present a serious CALC-driven problem in organization (high enough crew ratios), training (sufficient aircraft to train higher crew ratios), and equipment (enough airplanes to meet the CALC and training demands, not to mention MRCs).

The AWACS problem appears to have reached a head in the accidental downing of two U.S. helicopters over Northern Iraq in April 1994, with some of the errors being attributed to the AWACS crew. That tragic event was foreshadowed for the authors of this report in their visit to the regional CinC in March 1994, a month earlier. The heavy demands upon the AWACS crews had previously arisen in discussions with the Air Combat Command in January 1994. Mindful of those demands, the authors asked senior members of the CinC's staff if they thought the CinC might turn away some of the rising demands of CALCs in his theater. The respondents thought not; it was simply not wise nor part of the military ethos for commanders to turn down requests for their services.[6] When the authors asked the respondents to speculate on what would happen if the demands kept rising and the available forces kept declining, several offered the grim possibility that they "would stumble badly" in some operation before the world awakened to the problem.

---

[4]Michael A. Dornheim, "Fogleman to Stress 'Stability' After Deep Cuts," *Aviation Week and Space Technology*, November 7, 1994, pp. 28–29.

[5]Based on a conversation with Colonel Richard Meeboer, USAF, an Air Force planner, on December 1, 1994.

[6]The following anecdote was offered to illustrate the military ethos: The new skipper of a Navy frigate was aghast at discovering the poor condition of his ship's engineering plant and set about to put it right. Before he had a chance, however, he was ordered to sail with the fleet. He requested that his ship be excused until he could fix serious deficiencies in his engine room. He was summarily replaced as captain, and the ship immediately put to sea with a new skipper. Thus, do military officers come to understand what is expected of them.

CALC TRAINING OR EDUCATION?

CALCs may demand education more than training. The difference is more than semantic: Education provides the knowledge; training provides the skills. A medical intern and a brain surgeon may have the same education; but their skills are likely to be worlds apart. Training tends to evaporate and needs frequent refreshing, but education is generally more durable.

Most flight and fighting skills are gained and maintained through practice in the form of training; but CALC duties are more likely to alter how these military skills are to be applied than to introduce entirely new skills. Fliers and warriors believe that they must train (exercise) continually if they are to keep their flying and fighting skills honed to a sharp edge. By contrast, policemen, once trained at their academies, are likely to limit their training to physical fitness and occasional sessions at the pistol range. Much of their time spent at the police academies is in education—the law, human relations, and so forth—subjects that, once learned, will be slowly mastered in the field through long experience rather than by continual training. When fliers or warriors must perform policelike functions in CALCs, their need, like that of the policeman, is more for education than for training.

A similar comparison can be made for the demands upon military medical personnel in war and in some CALCs: Military surgeons need to keep practiced for treating battle wounds. To gain such continuing practice, they have resorted to working in emergency rooms in urban crime areas and to practicing surgery on animal carcasses subjected to projectile wounds. But in CALCs, military surgeons are more likely to be confronted with disease than with projectile wounds;[7] and in treating diseases, they rely more upon their education than practice. Nevertheless, their contributions in CALCs would probably be enhanced if they had more education on the diseases they would be most likely to encounter in CALCs—such as tropical infections or malnutrition.

---

[7]The Bosnian conflict is clearly different from most CALCs in that the casualties are more from the fighting than from disease.

As with basic lifesaving skills, such as the proper use of a tourniquet or CPR, most CALC skills—if they differ at all from routine military skills—are likely to be based upon understanding as opposed to practice. Projecting security from the air—air constabulary operations—is probably more a mindset than a skill. Once understood and reconciled with traditional flying and fighting skills, air constabulary operations may be less a matter of practice and more the application of other skills, such as airmanship, observation, discipline, and communications. Air constabulary skills for fighter, bomber, or gunship crews, once learned, are not likely to require continuing practice to keep them available for use.

Because the military supporting capabilities—logistics, transportation, security, and medical—are in the greatest demand for CALCs, they are likely to be easily applied without much additional training, practice, or education. Even for supporting units, however, there are some nuances found in CALCs that warrant education more than training. Since supporting personnel are more likely to come into direct contact with locals, language and cultural skills will always be in demand. For instance, in the USAF hospital in Zagreb, Croatia, staff are treating people from dozens of countries who speak a variety of languages. Because most of these patients do not have translators, the medical personnel communicate with them through sign language, observation, and touch.[8]

Developing those language and cultural skills may be less of a problem than accessing those skills quickly. The Air Force, as a large military institution, already has great diversity among its personnel in terms of language skills and cultural familiarity. Finding the person who has the particular background needed and making that person available quickly is an administrative challenge more than it is one of training or education.

For CALCs, more of the future airlifter missions may be expeditionary in nature—airlifters will sometimes be the first military mis-

---

[8]Based on conversations held at USEUCOM, March 7–8, 1994. Also see, for instance, William Mathews, "Surgery on the Spot," *Air Force Times*, April 17, 1995, p. 32, and Steve Salerno, "The Cutting Edge of Combat Medicine," *The American Legion*, March 1995, p. 20. It is also possible that cultural assistance may be transmitted through "telepresence" technologies.

sion into a locale and the crews will sometimes be on their own. Airlifters in the Cold War could expect to land at friendly, controlled airfields where they would be received, at the least, by U.S. military or State Department personnel. In CALCs, however, airlifters may be the first U.S. citizens in and the last out; and their single visit may be the entire mission. Like many CALC demands, being prepared to operate in an expeditionary mode is a mind-set rather than a separate set of skills. For airlifters, the mind-set must include the idea that they could be on their own—without ground-based air traffic control, landing area security, or crowd control—for cultural and language familiarity and for negotiations with locals for additional services or commodities, including the exchange of currency or credits.

Education, training and doctrine for CALCs must also address the likelihood of national interests or objectives changing during the course of CALC operations. War aims also have historically changed,[9] but "mission creep" in CALCs is a more sensitive issue for the U.S. military, partly because of the recent experience in Somalia, but more fundamentally because the U.S. military leaderships are less familiar and less comfortable with CALC objectives. Although the Air Force is not responsible for establishing CALC objectives, it does have a responsibility for *informing* civilian leaders of the limits and capabilities of air power and for anticipating the future needs for air power in CALCs. Therefore, USAF personnel must be educated to anticipate "mission creep" and must be prepared to inform their leaders not only of the dangers, risks, and costs, but of the military opportunities that may be presented by CALCs.

> We should do everything we can to let leadership know what we can do, but . . . [also what] the impact on our ability to fight will be.[10]

---

[9]For example, the war objectives during the Korean and Gulf wars underwent changes as opportunities and problems presented themselves.

[10]Admiral William Owens, Vice Chairman of the Joint Chiefs of Staff, as quoted in *Defense Daily*, August 20, 1994, p. 279, elipse in the original.

LEARNING FROM CALCS

CALC lessons learned on the job are more perishable than those learned in war or combat. The military leaderships and all of the forces watch, carefully document, and attempt to learn from combat experiences. In-depth studies are made after each war to reshape the concepts, doctrine, and training for the next. In the CALC air operations over Bosnia, the lessons learned in Operation Deny Flight (in which combat aircraft were employed in combat) are likely to be remembered long after the lessons learned in Operation Provide Promise (in which airlifters delivered or dropped supplies) are forgotten. The reasons for the differences in attention and memory are not hard to see:

- Combat and war are what the forces see as their primary purpose. Delivering supplies is seen as a means to that end, not as an independent end in itself.[11] So, even where the logistical efforts are more substantial than the combat operations, it is the combat operations that will be most remembered for lessons learned. The USAF leadership and much of the force will study the problems posed by Bosnian-Serb air defenses; but fewer will study the problems of making effective airdrops of supplies to the Bosnian Muslims.

- As in Somalia, CALC experiences are often limited to smaller portions of the force; and those experiences may be seen as mostly or only pertinent to that region. The language and cultural knowledge that the Air Force assembled to deal with Somalia will probably dissipate quickly because a repetition of that operation—in kind and locale—is thought unlikely.

Whether that selectivity of attention and memory is warranted by the Air Force's constant responsibilities and the changing nature of the world in which they will be applied is certainly arguable. Whether the Air Force should make any greater effort to collect and retain the learning gained from CALCs is no less arguable, for the answer depends further upon the perceived costs and benefits of the learning effort.

---

[11]Actually, delivering supplies over Bosnia became a *political* end in itself, as some involved in the airdrops will attest.

In a CALC-dominated world, however, it would seem, as a minimum, that personnel rotations should be monitored so that lessons learned from one CALC can be retained and mixed appropriately for subsequent CALCs. There is evidence that some lessons learned in one CALC have been applied to the next. One lesson learned from Operation Provide Hope and applied to Operation Support Hope was the need to set up a distribution system quickly, even before trying to make many food deliveries. The Kurdish relief efforts also taught the importance in Rwanda of establishing way stations along a refugee route, not only to provide assistance, but to entice migrants to return home. Operation Provide Promise taught the USAF the need for accurate and benign distribution of supplies in Operation Provide Hope.

The Air Mobility Command (AMC) has taken the first step toward reevaluating how its airlift capability should adapt to CALCs. AMC implemented a new initiative called "Realistic Training," which addresses the need to communicate the lessons learned in the field to the rest of the command. First, AMC ensured contact with and feedback from the field. Headquarters staff officers on flying status visit all AMC wings to interview personnel on their most recent CALC experiences. Second, Realistic Training provides a coherent link between AMC training and the need for additional education in CALC operations.[12] A USAF-wide program of feedback between planners and operators similar to the AMC program would seem to be a logical expansion for improving USAF operations in future CALCs.[13]

WHERE MRC TRAINING IS NOT ENOUGH

For the USAF, the two-MRC strategy places heavy emphasis on air power to quickly bolster allied forces and stem an enemy invasion while waiting for ground troops to arrive in the theater. That MRC emphasis was evident in a recent CALC, the crisis response[14] arising

---

[12]AMC News Service, August 1, 1994.

[13]The other U.S. military services have instituted similar programs. See Art Pine, "Deployments Take Toll on U.S. Military," *Los Angeles Times*, March 19, 1995, p. 1.

[14]This crisis response turned out to be a CALC, by the definition used here. If the crisis had escalated to war with Iraq, the crisis response would have been the lead-in

from the Iraqi military buildup near the Kuwaiti border in October 1994. The United States ordered the deployment of U.S. Air Force F-4s, F-15s, F-16s, A-10s, F-117s, F-111s, C-130s and F-111s, B-52s, KC-135s, and airborne warning and control system aircraft with about 7,600 USAF personnel joining some 3,900 airmen already in theater.

The cutting edge of those MRC-configured forces—bombers and fighters—are likely to have a more limited and different role, if any, in many CALCs. The traditional role of fighters or bombers in striking at enemy targets may, indeed, arise in CALCs, but most often in the form of punitive strikes—such as in the El Dorado Canyon strike and those conducted as a part of Operation Southern Watch—in which a protracted campaign of bombardment or strikes is not contemplated. In those instances, the premium qualities sought in the strikes are confidence in strike success and minimum collateral damage—two qualities that can be moderated in an extended air campaign against an enemy in an MRC. Thus, CALCs are more likely to draw upon the specialized forces—stealth, precision strike, SEAD, ECM, AWACS—for stunning displays of air power virtuosity than they are for the depth or quantities of air power that are available to the United States for waging war. Hence, there is a shade of difference in the training for these CALC missions: They demand planning for first-time success, for precision, and for pathbreaking rather than learning from doing and adapting to changing circumstances.

Again, the traditional role of fighters in MRCs to achieve air superiority may arise in CALCs, but more often in the form of preventing proscribed uses of the air by others. Instead of planning and waging an air superiority campaign involving successive attacks on enemy air defenses and air forces, CALCs may require that air power be used reactively, responding only to violations of rules imposed upon an adversary and then only under complex rules of engagement. These kinds of air superiority operations relinquish the initiative to the adversary and may often have to be conducted where innocent (e.g., civilian) uses of the airspace are allowed and ongoing. Must pilots be *trained* differently for such CALC operations, in which they relinquish the initiative to their enemy? The answer is not obvious to the

---

to an MRC, very much as Operation Desert Shield was the buildup for the MRC of Operation Desert Storm.

authors; but the question needs to be weighed by those commanders and pilots who have been confronted with the problem. Education on the differences would certainly seem to be warranted, even if regular training is not. One of those differences is the dominance of *political* concerns in CALCs versus the dominance of *military* concerns in MRCs.

AWACS personnel, whose job in an MRC is to monitor airspace and provide coordinating information for friendly air operations, have found CALCs more rather than less demanding. Under the wartime conditions and problems for which they train, the airspace is occupied by friends or enemies. CALCs present an airspace where all objects cannot be so neatly divided and where local procedures rather than U.S. or allied rules may apply.[15] AWACS crews rarely train or receive proper preparation for potential CALC airspace problems. Familiarity with local procedures often is lacking, and training and mission accomplishment suffer as a result. Proficiency is difficult to maintain when AWACS crews are deployed to so many places and when entire crews and even individuals are rotated often. Under these circumstances, it is not surprising to learn that these crewmen have a harder time adjusting to CALCs. Their CALC operations present greater demands and a mind-set different from that required by the MRC environment for which they have been trained.

SOME USAF CAPABILITIES ARE "JUST RIGHT" FOR CALCS

Even though designed for MRC environments, the applicability of airlifters and AWACS to CALCs is direct and obvious. But there are other, perhaps less obvious capabilities residing within the Air Force that can also be applied very effectively and directly to CALCs, not only in operations, but in the concepts, doctrine, and training for CALC operations. The largest of these capabilities is the Air Force Special Operations Forces (AFSOF), which have been designed around unconventional warfare skills for use in low-intensity con-

---

[15]For example, in Operation Provide Comfort, intended to protect the Kurds from the Iraqis, the Turks, who host the supporting AWACS operations, may be using the AWACS and other information to attack the Kurds in Turkey. See Vago Muradian, "Is U.S. Intelligence Being Misused?" *Air Force Times*, December 12, 1994, p. 20.

flicts (LIC)[16] and MRCs. AFSOF offers an ideal force in many CALCs because of its regional expertise in the Third World, extensive training in linguistic and cultural skills, specialized equipment designed for missions over all types of terrain, and its ability to deliver troops, equipment, and supplies onto short landing strips or small drop zones.

The spectrum of AFSOF activities during Operation Desert Shield and Desert Storm demonstrated special operations forces capabilities that are frequently in demand for CALCs: AFSOF provided recovery for coalition air forces in Iraq, Saudi Arabia, Kuwait, Turkey, and the Persian Gulf. AFSOF also provided emergency evacuation coverage for Navy sea, air, and land (SEAL) teams penetrating the Kuwaiti coast prior to the invasion. AFSOF demonstrated that it could conduct night operations near or in hostile territory with its full complement of planes and helicopters—AC-130s, HC-130s, MC-130s, MH-53s, and MH-60s.

Immediately after Operation Desert Storm, AFSOF went on to apply some of those capabilities in a CALC: Operation Provide Comfort in Northern Iraq. On April 14, 1991, MC-130E Combat Talons from the 7th Special Operations Squadron were the first to air-drop emergency assistance supplies to the Kurds in Northern Iraq.[17] Many of the refugee camp locations were in austere terrain and airdrops were the only means of survival until land routes and ground personnel could assume the humanitarian assistance effort. Clearly, participation in the Kurdish relief CALC served to sharpen AFSOF skills for future humanitarian missions and potential MRCs.

AFSOF personnel are very well suited to perform in many CALC missions because of their unique organization, training, and doctrine.

---

[16]Low-intensity conflicts may or may not be CALCs by the definition used here, depending upon whether or not the U.S. contributions have become routine, as in the long-term provision of advisors against a continuing insurgency in Latin America. A dynamic insurgency, with pending or actual significant changes in the U.S. commitments would be considered a CALC here. The long-term, low-level U.S. military assistance to the Peruvian government in dealing with the *Sendero Luminoso* or drug trafficking would not.

[17]John A. Hill, *Air Force Special Operations Forces: A Unique Application of Aerospace Power*, Maxwell Air Force Base, Alabama: Air University Press, April 1993, p. 10.

Their airlift capabilities for insertion, support, and extraction are of the kind frequently needed in CALCs for evacuation and rescues. Their facilities for PSYOPS (psychological operations) are more likely to be central to CALCs than to MRCs. Perhaps most pertinent to CALCs are these two assets:

- AFSOF personnel have extensive experience in recruiting, training, or utilizing large numbers of indigenous personnel and organizations that can be extremely beneficial in most CALCs.

- AFSOF has access to information in SOCRATES (Special Operations Command Research, Analysis and Threat Evaluation System), which incorporates a variety of computers, databases, intelligence communications systems, secure telephones, facsimile equipment, imagery processing and mapping services, and access to national and regional intelligence.[18]

Moreover, current SOF education programs could be a model for the expansion of a USAF-wide CALC program, with a majority of the syllabus devoted to education instead of training. The USAF Special Operations School catalog lists the types of classes that could be useful to USAF personnel as they enter or transfer to a new command where CALCs may occur. Specific orientation course materials on Latin America, the Middle East, Africa, and the Asia-Pacific region should be of assistance to airmen assigned to these regions. At a minimum, the USAF could consider distributing such materials to airmen as a basic introduction to their new AOR (area of responsibility).

Another USAF capability with direct and significant applicability to CALCs resides in the RED HORSE (Rapid Engineer Deployable, Heavy Operational Repair Squadron Teams) units for austere airfield operations. RED HORSE units are particularly well suited to CALC environments because they have been designed to be self-sustained with their own medics, cooks, and security forces. In Somalia, for instance, RED HORSE provided security for coalition forces from repeated rebel mortar attacks while performing their regular duties at Mogadishu Airport. The Somalia operations tested RED HORSE

---

[18]John M. Collins, *Special Operations Forces: An Assessment 1986–1993*, Washington, D.C.: CRS Report for Congress, July 1993, p. 25.

skills and demonstrated that they were well prepared to deal with conditions that are likely to be the norm in CALCs.

TRAINING IS NOT A LONG POLE

Training is not a long pole in the Air Force CALC tent. Training will not present any major problems for the USAF in readying personnel to perform in CALCS. To the contrary, without any dramatic changes in its current training—perhaps only with minor adjustments in specific educational programs—the USAF appears to be perfectly able to train or educate its personnel to serve alternatively as warriors, policemen, or social workers. Since the end of Operation Desert Storm in February 1991, most of the major operations mounted by the USAF have been humanitarian ones under the CALC umbrella—to save Kurds in northern Iraq (Operation Provide Comfort), feed the starving in Somalia (Operation Provide Relief), parachute and deliver food and medical supplies to noncombatants in Bosnia (Operation Provide Promise), and provide relief supplies to fleeing Rwandans (Operation Support Hope).

Although some training seems essential to an improved capability for CALCs, a hypothetical Air Force optimized for CALCs would probably be more differently educated rather than more differently trained.[19] Most of the skills required for CALCs are already found in an Air Force designed for MRCs, but they might be needed in different proportions. Flying skills are certainly no less needed, but fighting skills for CALCs are needed in less depth or numbers than for MRCs, and with an emphasis on responding to the immediacy of situations rather than on waging planned campaigns. It seems likely that many of the skills now found in Special Operations Forces, RED HORSE, and in the Air Police could be needed in greater numbers for CALCs. Spreading their existing knowledge and skills more widely through the Air Force may be the biggest training challenge posed by CALCs.

Because CALCs seem to require more education than training, the professional military education (PME) programs offered at the Air

---

[19]Training in the form of planning exercises is recommended in Carl H. Builder et al., *Report of a Workshop on Expanding U.S. Air Force Noncombat Mission Capabilities*, Santa Monica, Calif.: RAND, MR-246-AF, 1993, pp. 68–69, as a way to bring CALC problems to the attention of more people in the Air Force.

University may be one of the best places for the Air Force to look for opportunities to insert CALC considerations into the thinking of airmen. At the Air Command and Staff College, the important tactical and planning differences (e.g., rules of engagement, campaign planning) between CALCs and MRCs could be developed, while at the Air War College, the political and strategic differences (e.g., mission creep, nongovernmental resources) would be appropriate issues. The same suggestions are, of course, appropriate for the other services and for the joint PME programs, if not already in effect.

Chapter Five

EQUIPPING FOR CALCS

The current airlift dilemma is the product of past neglect by Air Force budget planners who favored fighters, bombers, and satellites, not transport planes and tankers.[1]

We need fighters with night and all weather capability. We need tankers, usually more than we thought. We need airlift and humanitarian resupply as well as the movement of supplies to our own forces. We need all of the capabilities to support these forces across the spectrum, of conflict—electronic combat, reconnaissance, and special forces. We also need bases from which to fly all of these sorties that support these operations.[2]

At the outset of this research, improvements in Air Force capabilities for CALCs—whether by organization, training, or equipment changes—were thought to reside mostly on the margins: small changes here and there ought to provide significant benefits. For the most part, that presumption appears to have withstood closer examination of CALCs and Air Force capabilities. But equipment is the one area in which USAF CALC capabilities might be considerably expanded, albeit at considerable cost. There are some equipment improvements for CALCs that could be made on the margin, but there are also some that would rank as major development and acquisition

---

[1]Then CinC, Air Mobility Command, and now Air Force Chief of Staff General Ronald R. Fogleman, as quoted in *Air Force Times*, August 29, 1994, pp. 12, 16.

[2]General Robert C. Oaks, "Regional Conflict Today: A European Perspective," *Aerospace Power: Regional Conflict in the 1990s*, Aerospace Education Foundation, 1994, p. 52.

programs. These major options for improving Air Force equipment for CALCs are taken up in the order of the authors' estimates of their contributions to CALC capabilities—that is, the likely result of their being needed and of their probable contribution if available for future CALCs. Considered are the abilities to accomplish the following from the air:

- Detect, locate, and immediately suppress heavy weapons fire

- Suppress open urban disorders, without resorting to lethal means

- Drop or deliver supplies with PGM (precision guided munitions) accuracy, without landing

- Unload and pick up on short notice a small combat team or equivalent cargo in any cleared area anywhere in the world, at any time, in any weather

- Deliver large quantities of inexpensive, lightweight, self-erectable, disposable housing and medical structures

- Locate nuclear materials on the ground, at least to the extent now possible with civilian aircraft.

COUNTERBATTERY CAPABILITIES

Air power should be able to effectively "nail the smoking gun"—to immediately engage and suppress heavy weapons fire. Current air power equipment and doctrine are designed for attacking artillery en masse, wherever and whenever it is detected, and with little concern for collateral damage. What is needed instead is reactive, directed counterbattery capabilities—the ability to return fire, round for round—from the air, without having to put forward air controllers on the ground where they can be turned into hostages.

Counterbattery capabilities against mortar and artillery fire do exist today. They take the form of "fire-finder" radars that are capable of tracking ballistic projectiles in flight and back to their sources and of aimed-fire weapons capable of quickly countering the offending "battery." Those counterbattery capabilities—both the radars and the aimed-fire weapons—now reside in ground units. When the targets threatened by fire from mortars or artillery are U.S. installations, such as U.S. airfields in Vietnam, it makes sense to have counterbattery capabilities vested in ground units collocated with the U.S. installations. But in situations such as in Bosnia, where the United States has no combat units on the ground—and would prefer not to put any there—then counterbattery capabilities are currently denied: The best that can be done is to rely on foreign observers on the ground to direct U.S. air strikes, typically using iron bombs, with results that are probably not worthy of the aircraft sorties, as measured by the probabilities of success, collateral damage, and risks of aircraft loss.

Counterbattery capabilities are likely to be a perennial need for CALCs. The death and destruction produced by heavy weapons, such as artillery and mortars, are significantly greater and less discriminating than those produced by small arms. The introduction of heavy weapons (and military vehicles) into a conflict is likely to be seen as a regrettable escalation of violence, one more likely to produce collateral damage and innocent victims, and therefore to be prevented or deterred, if possible. Of course, small arms also produce collateral damage or innocent victims. The differences between small arms and heavy weapons are both quantitative and qualitative:

- Heavy weapons are far more effective than small arms in producing death and destruction.

- Small arms have generally proven impossible to proscribe, even in well-regulated civil societies.

- The range of heavy weapons tends to isolate or detach their users from the effects of their use upon their targets or victims.

Hence, the proscription of heavy weapons is generally taken to be a more practical and effective avenue to the suppression of violence.

If counterbattery capabilities, even though desired for a peacemaking or peacekeeping mission, always require placing U.S. units on the ground, at risk, in the middle of someone else's conflict, then they will often be eschewed unless there are other, more compelling reasons to put U.S. forces on the ground. When they are desired, counterbattery capabilities from the air are effectively made hostages to hostile ground forces, because without putting observers—and potential hostages—on the ground, there can be no counterbattery capabilities! The way out of this conundrum, of course, is to create *independent* airborne counterbattery capabilities. And if air power *were* to have independent counterbattery capabilities, those capabilities might very well be useful in MRCs as well.

The technical possibility of combining fire-finder radars and aimed-fire weapons on the same airborne platform would seem to be an obvious solution. Putting the fire-finder radar on a rapidly moving aircraft certainly complicates the fire-control problem, but that should be solvable with modern computers. The aimed-fire weapons could be provided by guns of the type now carried by C-130 gunships. The authors have not investigated the technical issues or costs that might be involved in adapting C-130 gunships or alternative airborne platforms to counterbattery missions—nor would they be the best people to do so—but the possibilities ought to be explored by people who are competent to judge the feasibility of the concept.

One can only speculate how the Bosnian conflict might have evolved if the Air Force had been able to employ airborne counterbattery capabilities from the very beginning. If the conflict had been limited to small arms, it might not have been any easier to resolve the issue—indeed it might only have protracted the conflict—but the destructive attacks on the cities and the paralyzing attacks on the airfields might not have been the televised face of the war. Infantry wars are different from artillery wars in destructiveness, just as armored vehicle wars are different in maneuvers and the mobility of forces.

SUPPRESSING URBAN DISORDERS

The suppression of urban disorder is another perennial need for CALCs. The need has surfaced for U.S. forces in Panama, Somalia, and most recently, in Haiti. Urban areas provide the disorderly with an environment that is rich in both targets (for destruction or looting) and cover (for concealment or refuge). The ability to suppress urban violence quickly and humanely from the air could greatly relieve the burdens on, or even eliminate the need for, U.S. security forces on the ground where they may become hostages rather than security enforcers.[3]

Whether it is practically possible to suppress urban disorders from the air is a technical question whose answer pivots on developments currently found under the rubric of "nonlethal" weapons (NLWs).[4] In a draft directive, U.S. Defense Secretary William J. Perry stated that "Nonlethal weapons can make available significant new capabilities in some circumstances to achieve military objectives while minimizing fatalities and undesired damage to property and the environment."[5] The kinds of NLWs under consideration include lasers, microwaves, sound waves, strobe lights, electromagnetic pulses, microbes, chemicals, and giant nets. Other NLW technologies explore the possibilities for super caustic chemicals that could eat through metal or rubber or plastic (to disable not only tanks and trucks but virtually any machine) and sticky or slippery foams (to incapacitate vehicles or people).

The term "nonlethal" has been challenged on the grounds that few measures purporting to be nonlethal can avoid the possibility of lethal effects: sticky foams might suffocate;[6] slippery foams might result in lethal falls. Nevertheless, there are crowd and individual control devices that are generally considered to be nonlethal in most circumstances. These include tear gas, water cannons, rubber bullets, electrical shock devices, and pepper spray. In Somalia, troops used cayenne pepper spray as a means for applying proportional force against low-level threats. At times, merely waving the aerosol can in the air was sufficient to ward off Somalis.[7] To what extent

---

[3]An early and vocal advocate of the ability of air power to suppress urban disorders is Colonel John Warden, Commandant of the Air Command and Staff College.

[4]See *Asia-Pacific Defense Reporter*, April-May 1994, p. 36; Paul R. Evancoe, "Non-Lethal Alternatives Weighed By Law Officers," *National Defense*, May–June 1994, p. 28. Ed O'Connell and Tom Dillaplain discuss air power applications of NLWs in their article, "Nonlethal Concepts: Implications for Air Force Intelligence," *Airpower Journal*, Vol. 8, No. 4, Winter 1994, pp. 26–33.

[5]*Defense News*, September 19–25, 1994, p. 6.

[6]"War and Police," *Washington City Paper*, July 22, 1994, p. 17.

[7]Jonathan T. Dworken, *Rules of Engagement (ROE) for Humanitarian Intervention and Low-Intensity Conflict: Lessons from Restore Hope*, Alexandria, Va.: Center for Naval Analyses, 1993, pp. 18–19.

could such means be effectively adapted for airborne delivery or use to avoid putting security forces on the ground and, hence, at risk? The use of such agents from the air (at greater distances) would seem to increase the problem of identifying the targets as innocent or otherwise; but that problem also depends upon the risks such agents create for injury to, rather than, say, immobilization of their targets.

If effective nonlethal means for suppressing urban disorders from the air could be developed, those capabilities would bring air power to bear in many important situations in which the United States would like to help indigenous security forces but would prefer not to put U.S. citizens at risk on the ground. With such capabilities, fewer U.S. and Somali lives might have been lost in Mogadishu. With such capabilities, fewer U.S. security forces would have been required in Panama and Haiti as indigenous security forces were being reconstituted. Although such capabilities might remain beyond the resources available to most or all domestic law enforcement agencies, they would, nevertheless, be available to the U.S. president if law enforcement agencies and national guards were overwhelmed by widespread domestic disorders.[8]

PRECISION SUPPLY DELIVERY

Air-dropping supplies is a commonly employed air power capability in CALCs. Airdrops are more expensive and less efficient than alternative means of air and ground transport; but if air bases are not available close by and the distances are great, or if the ground environment is hostile, then airdrops may be the most expeditious and practical means for quickly getting supplies into the hands of those who need them. Airdrops were a common operation in the Vietnam War when isolated units needed supplies and the terrain or hostile

---

[8]Contrary to popular belief, Posse Comitatus does not prohibit the use of regular, active-duty military forces in domestic disorders; it only prohibits their being placed under any authority not in a direct line to the president as their commander in chief. National guards report to the authority of their state governors unless federalized; federal forces must report to the president. Posse Comitatus is not proscriptive of the military, but of civilians other than the president who might misuse the military. The penalties for violation of Posse Comitatus apply to any *civilian* who would attempt to break the chain of command to the president and exercise authority over federal military forces.

fire prevented landing aircraft or helicopters. Airdrops became newsworthy once again in the early relief efforts for the Kurds in the rugged terrain of northern Iraq, before surface and air base access to the region could be established. In Bosnia, airdrops became the means for supplying isolated enclaves of civilians when surface access was denied.

The long-standing problem with airdrops is precision—putting the supplies precisely where they are wanted, with confidence. Colonel John Warden, Commandant of the Air Command and Staff College at the Air University, has pointed out on numerous occasions that if air power can deliver munitions with high precision, it should also be able to deliver supplies—food or medicine—with the same precision. The need for precision in airdrops is the same as the need for precision in airdrops of munitions:

- To place the munitions or supplies where they are needed, with confidence, effective for the purposes for which they were carried and dropped.

- To avoid unintended or collateral damage from having the munitions or supplies land where their effect is contrary to the purposes for which they were carried and dropped.

- To conserve resources by ensuring that minimum possible amounts of munitions or supplies are carried and dropped to achieve the intended purposes.

In the humanitarian airdrops to the Kurds, in Operation Provide Comfort, and over Bosnia, as a part of Operation Provide Promise, all three reasons for precision were painfully evident: In the airdrops over Bosnia, some early drops missed their intended targets completely and were effectively lost to their intended recipients. A few of the early pallet drops to the Kurds resulted in injuries and death to the eager recipients; and some of the early food drops to Bosnian Muslims were reported to have ended up being dropped to the Bosnian Serbs.[9] In Rwanda, airdrops nearly hit a Zairian school and

---

[9]However, given the nature of the Bosnian conflict, it could be argued that food or medicine could hardly fall into the "wrong" hands if the purpose of the drops was humanitarian.

a U.N. helicopter.[10] In Bosnia, some of these problems were addressed by deliberately scattering food rations over a wide area; and although that improved the prospects of some food reaching its intended recipients and reduced the risks of unintended damage to structures, it was just as clearly an inefficient use of the resources.

The concepts available for precision drops of supplies are at least as numerous as those for delivering munitions, because some of the constraints on the delivery of munitions—high speed to avoid defenses, agility to hit moving targets, and so forth—do not necessarily apply to the delivery of supplies. Among the obvious alternatives to be considered are:

- Terminal guidance via laser designation or image-matching.

- Navigation via radar-tracking and command guidance or Global Positioning System (GPS) coordinates.

- Maneuvering via wings, fins, propulsion, parasails, and steerable parachutes.

The Army has long been concerned with more precise delivery of its supplies, and it plans to purchase the Advanced Precision Airborne Delivery System (APADS).

> APADS will use a packaged, nonrigid wing that extracts itself out of the back of a high-flying C-130 aircraft through use of a drogue parachute. Once the APADS platform stabilizes itself, it snaps into its gliding-wing configuration. From there, APADS' GPS (Global Positioning System) satellite navigation package guides the cargo to within meters of [a] predetermined landing point.... [T]he APADS glider can utilize a series of turns and maneuvers to deliver its payload up to 40 miles from the drop site.[11]

The concept is still in the research phase, but its advocates see that the advantages of a "drop and forget" precision delivery system will

---

[10]*Los Angeles Times*, July 29, 1994, p. 8. U.S. military officials stated that the drops were accurate since the U.N. provided the drop-site data.

[11]Eric Klingemann and Michael Seawood, "Anytime, Anywhere," *Armed Forces Journal International*, October 1994, p. 22. For more on APADS, see Sheila Foote, "Army Buying Remote Control System to Guide Airdrops," *Defense Daily*, March 9, 1995, p. 343.

apply to the Air Force as well as the Army: "Since it is designed to be dropped from higher altitudes and greater standoff distances than are currently used, the risk to Air Force pilots from enemy air defenses [should be] reduced." From the CALC perspective, this is not a capability whose future the Air Force should leave to the fate of Army interests in a time of declining budgets; it is a capability that could importantly affect the utility of air power far beyond Army missions.

GLOBAL TOUCH-AND-GO

Although air power has attained global reach, in large part through in-flight aerial refueling, it remains a potential hostage for ground forces in those areas where it must insert and extract people and other delicate cargoes in hostile environments. The *insertion* of trained people and some cargoes in hostile environments can be accomplished by parachute drop; but their *extraction* still depends almost entirely upon helicopters; and helicopters mean ranges that are far short of global. The disaster at Desert One in April 1980, the failed attempt to rescue the Iranian embassy hostages, highlighted a needed capability for air power in CALCs that remains unfulfilled after more than 15 years. To extract personnel during the recent Rwanda crisis, Marine helicopters were once again forced to fly for extremely long periods and to rely upon successive in-flight refuelings, just to span ranges from the coast to the interior of the African continent. Some things besides people cannot be air-dropped; they must be inserted, gently: In the Afghan war, a CALC before the end of the Cold War, the United States found it in its interests to transport livestock, fruit trees, and supplies to the Mujahedeen in Pakistan.[12]

It should be possible, with modern aerospace technology, to reach any site on the globe within 24 hours and to insert or extract, say, a dozen people and their baggage, with nothing more than a sports (i.e., soccer) field for a landing point. The ability to use any sports field of choice, day or night, and to insert and extract quickly (say,

---

[12]This event offers another lesson: Some CALCs can destroy indigenous animal transportation and food supplies. *MAC and the Afghan Humanitarian Relief Program,* Scott AFB: Military Airlift Command, undated, p. 2; David P. Masko and Michael James Haggerty, "Humanitarian Airlift, " *Airman Magazine,* February 1992, p. 17.

within 15 or 20 minutes) should greatly decrease the prospects of encountering prepared hostile actions. The global reach cannot be provided by a helicopter; and the sports field insertion and extraction cannot be provided by a long-range airplane. A combination of the two in a mother-daughter arrangement is one possible form. The daughter could be a helicopter, autogyro, or a VTOL (Harrier or Osprey-like) aircraft. Whatever the choice, the daughter aircraft need not have an operating radius of more than one or two hundred miles. The mother could be a turboprop or turbofan driven long-range aircraft with in-flight refueling capabilities, perhaps not unlike the designs considered by the Air Force during its LRCA (long-range combat aircraft) studies of 15 years ago.

The development of such a capability would be costly, measured in billions of dollars, even though the numbers of systems acquired would be small; perhaps no more than a squadron or two would be required. The expenditure of such enormous sums for such a small force makes the prospects remote, even if budgets were not declining. But until the Air Force has the capability for "global touch and go," global power will not include the power to insert and extract people anytime, anywhere, with a global reach. Its ability to insert and extract at global range will remain vulnerable to airfields and hence to easy military countermeasures.

INSTANT HOUSING

Housing is a perennial problem in CALCs, especially but not exclusively in relief missions. Disasters, whether natural or man-made, frequently cause homelessness: Either houses are destroyed or the inhabitants are forced to leave their houses to seek relief elsewhere. Recent CALCs have included the temporary housing of Kurds, Bosnians, Somalians, Rwandans, Cubans, and Haitians.

The military solution to temporary housing is tents. Although tents are a solution, they are by no means the best one for CALCs. Tents are an Army solution for the housing of soldiers in the field for a campaign. As such, they are:

- Erected, taken down, and moved by soldiers in the field; they are soldier's mobile homes, well-designed for a campaign of months and for transport by trucks.

- Manpower intensive for their erection and take-down—in keeping with their use in a manpower-rich field Army in which soldiers can assume responsibility for setting up their own mobile homes under the discipline and basic field skills of an army.

- Not designed for air transport, even though they may be air transported to the field; they are bulky enough to require teams of people to handle them and trucks to move them.

- Constructed with materials and concepts that are more than a century old; indeed, some used in recent CALCs were fabricated about a half a century ago.[13]

If the CALC task is to temporarily house civilian refugees or victims, the use of army field tents amounts to adapting a very old military solution to a problem that fairly begs for a modern, high-technology solution, especially if that housing must be brought in by air. Tenting technology has made great progress in the past half century, judging by the diversity of tent designs and materials available for mountaineering and other outdoor sporting activities. But even better housing solutions may be available in space-age composite materials—such as paper honeycombs, foils, plastics, foams, fiberglass, and fiberboard—formed into self-erecting shelters that fold flat, weigh very little, and can be popped into shape with cartoonlike instructions. Such modern collapsible housing is probably ideally suited for air transport in quantities that far exceed conventional army tenting. A family shelter can almost certainly be carried by a single individual and set up by that individual without any assistance beyond that of looking at a cartoon or watching others "kick" their package into a "pop-up" shelter.

Army tents, however, like military rations, are "free" goods, originally purchased for other military reasons. Why, then, should the U.S. military pay out of its pocket the price for space-age housing to be given to refugees or victims in CALCs? The answers are several:

---

[13]Although not a CALC example, the army tents employed in the aftermath of the Northridge (Los Angeles) earthquake in January 1994 were reported to be of Korean War vintage. They leaked during the rains that followed and had to be covered with more modern plastic sheets to make them waterproof.

- Army tents are in fact not free any more than are MREs (meals ready to eat); tents must be replaced when excess stocks are exhausted (in many CALC situations, tents are typically not reused after being distributed, because they have been subjected to soil, diseases, and misuse).

- Army tents require significant numbers of military personnel to remain on the ground for their movement and setup, making military personnel vulnerable to subsequent events on the ground. It may be possible to air-drop emergency housing without putting any personnel on the ground.

- Army tents require considerably more airlifters committed to housing missions than would be necessary for transporting modern shelter systems.

If air power is to provide independent means for exploiting the air to help people in emergencies and to avoid vulnerability on the ground, then emergency housing designed specifically for air transport is one of the first and most obvious places to look for opportunities. Should the Air Force buy emergency housing out of its own pocket? It may not be necessary.

When it became apparent that MREs presented dietary problems for some ethnic groups, special rations were formulated and purchased by DoD for emergency distribution. If the Air Force were to identify concepts for making emergency housing far more efficient in airlifter use, far less demanding of personnel on the ground, and much faster to be deployed and employed, it seems likely that DoD would be interested in pursuing those concepts. Anything that enhances the ability of air power to contribute significantly and independently to CALCs would seem to be worthy of Air Force consideration.

This concept of efficiency and practicability could be expanded to cover the particular kind of medical facilities that are required in many CALCs. Few CALCs have required surgical facilities (Bosnia is one exception). The greatest need in most CALCs is for outpatient clinics and dispensaries to deal with malnutrition, disease, immunization, and so forth. The housing requirements for such clinics are much more modest than for battle surgical hospitals; and concepts similar to those urged here for emergency housing might be adapted. A medical clinic and its equipment might be reduced to airlifted or air-dropped packages that a few medical personnel could hand-carry and set up.

The USAF certainly has the capability for getting large hospitals to an MRC where large numbers of surgical casualties are expected. In October 1993, for example, the first USAF unit to get a new "chemically hardened" air-transportable hospital became operational. The hospital, when set up, covers 15,000 square feet and includes 25-bed wards, an operating room, a pharmacy, a supply room, an emergency room, and laboratory and X-ray departments. Seven C-141s are needed to transport the hospital and medical supplies.[14] Clearly, this USAF hospital has an impressive capability for an MRC or for those few CALCs in which surgical casualties are expected; but this is an MRC solution that is "over-designed" for the problems and needs that can emerge in most humanitarian disasters.

Moreover, in the future, the USAF may be called upon to deliver and render medical aid without the benefit of relying on the other services. Usually, the Air Force relies on Army and Marine personnel to set up the camps and medical facilities (such as a MASH)[15] and lacks the personnel to establish the facilities themselves. In Zagreb, for instance, Air Force medical personnel worked in a MASH facility rather than taking their own air-transportable hospital.[16] In war and in many joint operations, the fact that the USAF does not have the personnel to establish its own hospital units may not be a problem; but in CALCs there may be circumstances where *no* U.S. personnel should be left on the ground. In such cases, the Air Force will still be expected to deliver the needed facilities to civilian or other military practitioners on the ground. The development of such facilities

---

[14]*Air Force Times*, December 13, 1993, p. 8.

[15]In Operation Desert Storm, the medical and evacuation units requested by USCENTCOM and provided by the Air Force would not have been sufficient to handle the large number of predicted casualties. Furthermore, even though the units had to treat fewer casualties than were predicted, the units still experienced difficulty accomplishing their mission. Deployment units did not have enough or the right mix of personnel; supplies were often missing or outdated or were incompatible with available equipment; many personnel were not appropriately trained; and the system used to regulate the movement of patients did not function adequately. See *Operation Desert Storm: Problems with Air Force Medical Readiness*, Washington, D.C.: United States General Accounting Office, GAO/NSIAD-94-58, December 1993, p. 2.

[16]*Air Force Times*, October 4, 1993, p. 11.

should be based upon CALC and airlift or airdrop considerations, not upon the requirements posed by MRCs.[17]

NUCLEAR DETECTION

The detection and verification of nuclear materials or weapons is likely to be an increasing national security problem in the decades ahead, despite nuclear nonproliferation efforts by the international community. The collapse of the Soviet Union and the increasing availability of nuclear materials in commerce—both lawful and illicit—makes the prospects for nonnational and clandestine nuclear weapons much higher than they were during the Cold War. Some CALCs in the future may pivot on claims or suspicions about the possession of nuclear materials or weapons.

The U.S. domestic capabilities to detect and verify claims or suspicions about the location of nuclear materials in the United States are the responsibility of the Department of Energy (DOE), assisted by the federal and local law enforcement agencies. An important part of those capabilities is vested with the Nuclear Emergency Search Team (NEST),[18] with airborne units staffed and maintained by civilian companies under contract to DOE. The NEST capabilities include civil aircraft fitted with nuclear detectors that can be employed in area searches to locate suspected nuclear materials. The capabilities of these systems are, of course, sensitive information and subject to change with technical developments. It is enough, here, to note that the search capabilities are judged to be of sufficient utility to maintain them on standby for use in domestic emergencies.

Those domestic capabilities for nuclear detection from the air are not mirrored in military capabilities for international situations. The reasons for that gap are not all obvious: Domestic security may come before international security in priority, but not to the exclusion of international security capabilities. The domestic capabilities for nu-

---

[17]It is possible that reforms in medical evacuation procedures may be a first step in correcting the deficiencies. See James Kitfield, "A Bigger Job for Medevac," *Air Force Magazine*, March 1995, p. 52.

[18]The team has fewer than 40 full-time employees, but when there are alerts it relies on more than 800 nuclear experts who work at the Lawrence Livermore, Sandia, and Los Alamos national weapons laboratories.

clear detection cannot be confidently applied to international situations because both the operators and the aircraft are civilian: The flight crews could refuse commitments to dangerous international environments and the aircraft may be unsuitable in range and equipment. The absence of military capabilities for nuclear detection from the air—corresponding to those now being maintained by civilians for domestic service—may have simply dropped through the cracks in planning military missions.

Detection from the air of nuclear materials on the surface is technically more challenging than detection from the surface. The distance between the sensor and the nuclear materials is a first-order determinant of the problem. The reasons for resorting to nuclear detection from airborne sensors are the same as those in many search problems: access and speed. Airborne sensors can cover areas that may not be easily accessible on the surface and can sweep large areas quickly.[19]

For the USAF, the mission of detecting nuclear materials from the air may beg the harder military questions: What does or can one do if nuclear materials are detected? What good does it do to know where nuclear materials are located if the materials cannot be confidently captured or destroyed? Or, can the detection capabilities be countermeasured by appropriately shielding the nuclear materials? Such questions, although pertinent, reflect the "worst case" military-planning paradigm more than they illuminate the problem. The same questions arise for the domestic problem confronting DOE and the law enforcement agencies. Some answers are:

- Sometimes knowing is an adequate basis for a decision not to act. Nuclear hoaxes have been unmasked by searches for nuclear materials. Political leaders must sometimes prove, to the best of human abilities, that threats—such as bomb threats in buildings or vehicles—are no more than that.

- Where threats are found to be real, neutralization is never confidently assured. Bombs can be booby-trapped, and bomb-dis-

---

[19]Even so, the sweep widths from the air can be narrow enough to require a tight search pattern, similar to those required in airborne searches for submarines with magnetic anomaly detection (MAD) sensors.

posal teams have sacrificed their lives and limbs in unsuccessful attempts to defuse bombs. Insistence upon high-confidence means for effectively neutralizing threats is a modern U.S. national security planning mind-set;[20] law enforcement and special forces are accustomed to greater uncertainties in mission success.

• Nuclear detection, like all detection processes, can be countermeasured; but, like all countermeasures, such processes impose costs and limits upon those who seek to avoid detection. Metal detectors and baggage X-rays can be countermeasured by those who seek to smuggle contraband, but only at a price that would deter most individuals who might otherwise elect to smuggle.

The kind of scenario that might call for USAF capabilities for nuclear materials detection from the air has been described elsewhere[21] and can be synopsized as follows:

Without warning, a nuclear device detonates in Tel Aviv and responsibility is claimed by several shadow terrorist organizations, some of which have never surfaced before. A claim is made that another such device is in place in Haifa. Among other things, Israel asks for technical assistance from the United States to try to verify the claim. The risks for nuclear search teams, on the ground or in the air, are obvious.

The use of civilian contractor search teams in this kind of scenario seems improbable. What the Air Force would need to extend the airborne nuclear detection capabilities—that now reside with civilian contractors and aircraft—into international, military environments are military aircraft fitted with equivalent detectors and crews trained for searches under high-risk international conditions. The needed numbers of such USAF aircraft and crews would be compa-

---

[20]That mind-set probably reached its peak during the Cold War with attempts to ensure the pre-launch survivability of the strategic nuclear forces. No threat was considered too extreme, and no proposed basing scheme for missiles or bombers was excused from passing the test of such threats. The analyses of electromagnetic pulse (EMP), depressed missile trajectories, and exquisitely timed and coordinated massive nuclear attacks are testimony to the penchant for high-confidence neutralization of threats.

[21]Carl Builder et al., *Report of a Workshop*, pp. 33–40.

rable to those maintained by the civilian contractors for DOE, probably fewer than five. As such, they might be logically incorporated in the AFSOF units.

Although the need to employ military nuclear search teams may be a very low-probability event, the stakes could be very high; and the recriminations for failing to have undertaken modest preparations against such events—within the known state of the art—would be painful. These are sufficient reasons for maintaining the domestic capability for nuclear search, but are they sufficient reasons to field the same capabilities for international operations? The Air Force has probably undertaken more expensive preparations against arguably less likely events of arguably less consequence.[22]

OTHER EQUIPMENT NEEDS FOR CALCS

The above six capabilities do not, of course represent all of the Air Force capabilities required or desired for CALCs. Moreover, most of the capabilities that the Air Force brings to CALCs already exist; and, for the most part, no new equipment is required when those capabilities are applied to CALCs rather than to MRCs. The six capabilities described here represent those that do not currently exist and probably will not be developed for MRCs. These are capabilities that would enhance the application of air power to CALCs more than they would to MRCs, although several of them would probably be found useful if available for MRCs as well. Not included in these six capabilities are:

- Equipment that could be purchased off the shelf, such as aircraft with the ability to land small loads (e.g., a pickup truck) on primitive high-altitude strips.

- The modifications that might be made to existing equipment to improve their utility in CALCs—e.g., protection schemes for all airlifters and their crews against CALC threats such as small-arms fire and man-portable surface-to-air missiles (as opposed to sophisticated and extensive air defense systems).

---

[22]Several retrospective examples might be found in the Cold War measures to protect nuclear forces against all sorts of arcane contingencies. See fn. 20, above.

• Better means for stopping surreptitious flights by occasional low and slow flyers while enforcing air embargoes. Our current MRC-oriented equipment and doctrine are designed for massive attacks on enemy aircraft wherever they are—on the ground and in the air. But the rules of engagement for air embargoes may prohibit engaging aircraft on the ground. That means offending helicopters and light planes can *squat* on the ground when detected, to avoid being engaged.[23] If the embargo is enforced only with *fast movers* of limited flight endurance, the violators can simply out-wait the enforcers and then move on. The enforcing aircraft must also be able to squat and wait or, better yet, to squat and capture. This should be possible with helicopters and VTOL aircraft.

• Simple *give-away* communications and navigation gear (e.g., cellular telephones and GPS repeaters) that can be activated by USAF aircraft operating in the local area. Such distributed devices could be useful in identifying and localizing pertinent conditions on the ground that bear upon the effectiveness or execution of CALC operations.

• The rich and rapidly growing domain of *information warfare* as it may apply to CALCs when employed from air and space platforms. The authors have been wary of opening this box because its contents, like *nonlethal weapons,* are still formative and deserve separate and much more detailed discussion. Nevertheless, information warfare, even in its current forms, could turn out to be a more important capability in CALCs and in MRCs.[24]

Perhaps even more important, but not addressed here, is the need to rebalance existing USAF equipment inventories in a hypothetical

---

[23]It might first seem that the rules of engagement are at fault; but this is a commonplace police situation: Force may be used to halt actions that break the law, but if those unlawful actions cease, force may not be used against the lawbreaker simply because the law had been broken previously. Flying in a no-fly zone is unlawful; squatting on the ground is not. In the air, the violator can be attacked. Once on the ground, the violator can be guarded or captured but not attacked.

[24]The recent military intervention in Haiti involved Air Force C-130 aircraft broadcasting information directly to the Haitian populace as one of several means employed to avoid unnecessary conflict between the Haitians and U.S. forces.

military planning world dominated by CALCs instead of MRCs. It is apparent that on-going CALC operations already point to the need for more AWACS, SEAD, tactical and strategic airlifters, reconnaissance assets, and so forth. Just how those force structure balances should be struck now or in the future is a broader question than the one addressed in this research. The narrower question addressed here is, "What stands out as requiring attention?" "Which poles are the long poles in the CALC tent for the Air Force?" New kinds of equipment or capabilities described in this chapter could be some of those long poles. But changing the *force balance* or *force structure* to improve CALC capabilities is not just an Air Force CALC issue, it is a national security planning question.

Chapter Six

CONCLUDING OBSERVATIONS

Thinking about an Air Force designed for CALCs is only a reference point for thinking about what, if anything, should be done now. The most urgent problem caused by CALCs is relieving the stresses now falling on certain people, units, and equipment. Unrelieved, these stresses are causing critical people (along with their skills) to leave the Air Force, causing premature wear on critical equipment needed to prosecute MRCs, and setting up the Air Force for failure one way or another. Resources need to be redistributed insofar as possible between fighting and supporting units and between active and reserve units to relieve these stresses.

Beyond that, the Air Force needs to plan how it might respond in the future as the world evolves into one dominated either by MRCs or CALCs. The difference between the two orientations now lies on a knife-edge: A single MRC could strongly reinforce the current planning paradigm that focuses on MRCs. But a peaceful reunification of the Korean Peninsula or favorable changes in Persian Gulf regimes (and both possibilities appear in current regional speculations about the future) could make MRCs seem much more remote if the United States remains heavily committed to a number of CALCs. What the Air Force needs right now in CALC planning is not more money but more thought. Although many military leaders consider CALCs important—as evidenced by the quotations throughout this report—there are

- no institutional or bureaucratic pressures on the Air Force to even think about how it would realign its capabilities toward

CALCs at the expense of those for MRCs, if the need should arise, and

- no points of advocacy within the Air Force for CALC planning concepts—perhaps because such advocacy could be perceived as having the potential to produce additional, unwanted pressures on scarce resources.

But fears of budget pressures should not prevent the Air Force from thinking now about what kinds of actions would be prudent if CALCs should continue to grow in number and in scope and then begin to dominate Air Force operations of the future. The easiest and most effective way to engage Air Force thinking about CALCs would be to create a point for planning advocacy within the Air Staff at USAF Headquarters, in which the needed breadth of knowledge and expertise about the Air Force could be credibly connected to the challenges and opportunities for air power in CALCs.

TENSIONS POSED BY CALCS

However, taking that first step—and thinking about a future in which CALCs could be of ascending importance—is made more difficult by the tensions raised by CALCs for all U.S. military institutions. At least four such tensions can be identified and characterized by the "ghosts" they evoke:

1. A tension in the emphasis and priorities between maintaining force readiness and infrastructure or between force training and operations. CALCs would drive the emphasis toward infrastructure and operations rather than toward readiness and training. The ghost evoked by that tension is "Task Force Smith," the name given to the initial defense of South Korea in 1950, when ill-prepared forces were committed to battle with disastrous results. A rallying cry for readiness heard today, especially in the Army, which suffered the embarrassment, is "No more Task Force Smiths!" The impetus for the ghost, however, may have more recent origins:

> Poor readiness, many Army officers believe, tends to spread like an infection. Today's Army leaders were all midlevel commanders in the 1970s, and experienced firsthand that era's 'hollow Army'—a

dispirited, poorly trained force, where drug abuse and insubordination were rampant.

'There's a latent institutional uneasiness,' said one senior Army officer. 'People of my generation feel strongly about this because we went through a time when we had a very, very bad Army.'[1]

2. A tension in the purpose and identity of the U.S. military between warfighting and broad military service or between finite missions and open-ended tasking. CALCs would drive the armed services toward purposes associated with broad military service to the nation rather than toward "fighting and winning the nation's wars" and toward messy tasks at the political-military interfaces rather than toward well-defined military missions. The ghosts evoked by that tension are found in the "Coup of 2012"[2] and the mission creep to which many attribute the failure in Somalia.

3. A tension in budgets and resources (e.g., personnel ceilings) between expenditures on weapons and their support or between maintaining the size of the forces and the diversity of the skills in the forces. CALCs would drive the force structures toward a larger ratio of supporting elements and keeping a diversity of capabilities that are not considered mainstream by the armed services. The ghost evoked by that tension is found in the warning that the U.S. military could find itself, once again, relegated to "a few dusty camps in Kansas."[3]

4. A tension in the leadership and control of the U.S. military institutions between the operators and the logisticians or between teeth and tail. CALCs would push the balance of power and control over

---

[1] John F. Harris, "Military Readiness Question," p. 16.

[2] The term derives from a prizewinning term paper, subsequently published as an article, describing a scenario in which the assignment of noncombat missions to the U.S. military results in its being drawn increasingly into domestic affairs and politics, with the ultimate consequence of a military coup. The paper was praised by some as a well-articulated description of the potential dangers of assigning the military the "wrong" kinds of missions. See Charles J. Dunlap, Jr., "The Origins of the American Military Coup of 2012," in *Parameters*, Vol. 22, No. 4, Winter 1992–1993, pp. 2–20.

[3] The phrase is found twice in John Setear et al., *The Army in a Changing World: The Role of Organizational Vision*, Santa Monica, Calif.: RAND, R-3882-A, June 1990. It may refer to the poor state of the Army between the two world wars, when some could bitterly observe that the Army had been relegated to a few dusty camps in Kansas.

the armed services toward the logisticians and supporting elements rather than keeping it with the operators and the warriors. Indeed, CALCs raise the question of whether the armed services are to be mostly warriors, police, or social workers. The ghost evoked by that tension is that the "desk jockeys" will once again take over the military, as they are perceived to have done so many times before in peacetime.

Thus, if CALCs should continue to ascend in frequency or political importance, they will challenge some very basic institutional values of the U.S. military. Of the four tensions listed above, the last is the one that will probably dominate the resolution. The current military leaderships, forged during nearly 50 years of war—cold or hot—are almost exclusively drawn from the warriors and operators. They will naturally resist the changes that would shift the power or control in the institution toward those who have been in supporting roles.

A COMPASS FOR THE FUTURE

If the considerable hurdles—the resources, organization, and traditions that now impede thinking about CALCs as an important aspect of the Air Force's future—can somehow be overcome, what then? What concepts, strategies, or doctrines should guide the Air Force as it proceeds to organize, train, and equip forces for CALCs as well as for MRCs? One principle stands out from this research: CALCs can be quagmires. If air power is to offer a significant military alternative for the nation's leadership, it must not be held hostage by having to put people—airmen or soldiers—in harm's way, just to support air operations. Air power must provide independent means for doing the things that must be done in CALCs without committing people to the ground, even in supporting roles. This is not the traditional call for the independence of air power from ground commanders; it is a call for air power to give the nation's leadership an alternative that does not trap the nation in someone else's conflict. Air power must be able to feed, supply, rescue, police, and punish from the air, without resort to air bases within the afflicted area.

This challenge for air power is less technical than financial; and it is less financial than institutional: If the institutional Air Force makes up its mind to pursue such independent capabilities for air power in CALCs, the resources can be found. And if the resources are found, even in an era of sharply constrained budgets, the technical problems can be solved.

This challenge for air power is not unfamiliar. It is closely related to the challenges for air power that arose in the aftermaths of the two world wars: After World War I, the challenge for air power was to offer the nation's leadership a military alternative to the stalemated carnage of trench warfare on the ground. Air power offered the promise of leaping over those trenches and striking at the heart of the enemy—a way to avoid the bloody ground warfare that had cost the Europeans a generation of young men.

After World War II, the challenge for air power was to offer the nation's leadership a military alternative to ground warfare against hordes of soldiers that the United States could not hope to match in numbers. Air power, then pumped up with nuclear weapons, again offered the promise of leaping over the masses of soldiers and striking at the heart of the enemy—a way to avoid the kind of attrition warfare on the ground that the nation could not hope to win.

The pattern is evident: After each world war, air power developed and evolved by responding to the challenge posed not so much by the *next* war as evoked by the nation's *nightmare* of the last war. Today, the nation's nightmare does not seem to be of an MRC; instead, the MRC may be the U.S. military's standard for a "proper" war that can be fought and won. The nation's nightmare seems to be about finding itself held hostage—as it was in Vietnam, as the Soviets were in Afghanistan, as the Russians may become in Chechnya—in an endless, unwinnable conflict.

Now, after the Cold War, the challenge for air power could very well be to offer the nation's leadership military alternatives to crises and lesser conflicts that the nation wants neither to ignore nor to be entrapped by. Air power—with independent capabilities to feed, supply, rescue, police, and punish from the air—could, we think, be fashioned to address urgent problems without being held hostage on the ground. To meet that challenge, air power must be able to:

- Strike offending targets (e.g., heavy weapons) while they are in the act of offending, not later and not with controllers on the ground.

- Suppress open disorders from the air and with fewer injuries or deaths than with alternative ground-based means.

- Deliver supplies precisely, including housing and medical facilities, without the necessity for using air fields within the afflicted area.

- Put down and pick up small numbers of people at any time and at almost any place, without warning.

These objectives would seem to be within the reach of an Air Force that has proven itself capable of girdling the globe with its air and space vehicles, of turning the ground and air beneath its instruments (and all who move therein) into instant electronic maps, of making its aircraft nearly invisible to radar, and of many other things equally wondrous.

The challenge is there. So, probably, are the means, both technical and financial. But the challenge may not seem worthy of the costs— costs now measured mostly in what the institution has come to value—in traditional forces. The future development and evolution of air power could be in the balance. It has been before, in the 1930s, when the Army leadership thought that air power should be a service rather than a force. It was once again, in the late 1940s, when the Army and Navy leaderships thought that air power should not be independent from their surface forces. It may be now, over the relevance of air power to a world in which regular warfare seems less likely or less frustrating than the disorders and human tragedies that are increasingly emerging everywhere.

REFERENCES

Adelman, Ken, "Dialing 911 for the Military," *Washington Times*, August 12, 1994, p. 19.

Air Force Times, October 4, 1993, p. 11.

Air Force Times, December 13, 1993, p. 8.

AMC News Service, August 1, 1994.

Asia-Pacific Defense Reporter, April–May 1994, p. 36.

Builder, Carl H., "Looking in All the Wrong Places?" *Armed Forces Journal International*, May 1995, pp. 38–39.

Builder, Carl H., "Nontraditional Military Missions," in Charles F. Hermann, ed., *1994 American Defense Annual*, New York: Lexington Books, 1994.

Builder, Carl H., et al., *Report of a Workshop on Expanding U.S. Air Force Noncombat Mission Capabilities*, Santa Monica, Calif.: RAND, MR-246-AF, 1993.

Capaccio, Tony, "Supreme Allied Commander Sketches Challenges of 'New NATO,'" *Defense Week*, Vol. 16, No., 6, February 6, 1995, p. 8.

Collins, John M., *Special Operations Forces: An Assessment 1986–1993*, Washington, D.C.: CRS Report for Congress, July 1993, p. 25.

Compart, Andrew, "Decade of Change," *Air Force Times*, October 18, 1993, p. 20.

Compart, Andrew, "How Capable Are Fighter Units?" *Air Force Times*, May 30, 1994, p. 20.

Compart, Andrew, "Job Can't Be Done Without Reserves," *Air Force Times*, August 14, 1994, p. 13.

Compart, Andrew, "When Part Time Stops Being Part Time," *Air Force Times*, May 30, 1944, p. 20.

Defense Daily, August 20, 1994, p. 279.

Defense Daily, September 2, 1994, p. 354.

Defense News, September 19–25, 1994, p. 6.

Dixon, Anne M., "The Whats and Whys of Coalitions," *Joint Force Quarterly*, Winter 1993–1994.

Dornheim, Michael, "Fogleman to Stress 'Stability' After Deep Cuts," *Aviation Week & Space Technology*, November 7, 1994, pp. 28–29.

Dunlap, Charles J., Jr., "The Origins of the American Military Coup of 2012," *Parameters*, Vol. 22, No. 4, Winter 1992–1993, pp. 2–10.

Dworken, Jonathan T., *Rules of Engagement (ROE) for Humanitarian Intervention and Low-Intensity Conflict: Lessons from Restore Hope*, Alexandria, Va.: Center for Naval Analyses, 1993.

Editorial, "Military Isn't Dollar-Short," *USA Today*, November 21, 1994, p. 10.

Evancoe, Paul R., "Non-Lethal Alternatives Weighed by Law Officers," *National Defense*, May–June 1994, p. 28.

Fogleman, General Ronald R., Air Force News Service, October 12, 1994.

Fogleman, General Ronald. R., *Air Force Times*, August 29, 1994, pp. 12, 16.

Fogleman, General Ronald R., "Core Competencies—New Missions: The Air Force in Operations Other than War," presented at the American Defense Preparedness Association Symposium, Washington, D.C., December 15, 1994, as reported in *Air Force Update*, 95-01.

Foote, Sheila, "Army Buying Remote Control System to Guide Airdrops," *Defense Daily*, March 9, 1995, p. 343.

Gertz, Bill, "Clinton OKs Call-up of Reserves," *Washington Times*, September 16, 1994, p. 16.

Graham, Bradley, "New Twist for U.S. Troops: Peace Maneuvers," *Washington Post*, August 15, 1994, pp. A1, A8.

Graham, Bradley, "Pentagon Officials Worry Aid Missions Will Sap Military Strength," *Washington Post*, July 29, 1994, p. 29.

Harris, John F., "Military Readiness Question Is Founded in Debate over Roles, *Washington Post*, November 18, 1994, p. 16.

Hill, John A., *Air Force Special Operations Forces: A Unique Application of Aerospace Power*, Maxwell Air Force Base, Alabama: Air University Press, 1993.

Hoffman, Bruce, *British Air Power in Peripheral Conflict, 1919–1976*, Santa Monica, Calif.: RAND, R-3749-AF, 1989.

House, Karen Elliott, "The Wrong Mission," *Wall Street Journal*, September 8, 1994, p. 18.

Inside the Army, August 15, 1994.

Kitfield, James, "A Bigger Job for Medevac," *Air Force Magazine*, March 1994, p. 52.

Klingemann, Eric, and Michael Seawood, "Anytime, Anywhere," *Armed Forces Journal International*, October 1994, p. 22.

Linn, Thomas C., "The Cutting Edge of Unified Actions," *Joint Force Quarterly*, Winter 1993–1994, pp. 34–39.

Los Angeles Times, July 29, 1994, p. 8.

MacFarland, Margo, "Rethinking 'Tooth-to-Tail,'" *Armed Forces Journal International*, September 1994, p. 45.

Mahlburg, Bob, "Shalikashvili Says Aircraft Funding Among Tough Choices," *Fort Worth Star-Telegram*, August 25, 1994, p. 27.

Masko, David P., and Michael James Haggerty, "Humanitarian Airlift," *Airman Magazine*, February 1992, p. 17.

Mathews, Willliam, "Surgery on the Spot," *Air Force Times*, April 17, 1995, p. 32.

Maze, Rick, "Units Wield Political Clout," *Air Force Times*, October 18, 1993, p. 28.

Military Airlift Command, *MAC and the Afghan Humanitarian Relief Program*, Scott AFB: Military Airlift Command, undated, p. 2.

Muradian, Vago, "Is U.S. Intelligence Being Misused?" *Air Force Times*, December 12, 1994, p. 20.

Oaks, General Robert C., "Regional Conflict Today: A European Perspective," *Aerospace Power: Regional Conflict in the 1990s*, Aerospace Education Foundation, 1994, p. 52.

O'Connell, Ed, and Tom Dillaplain, "Nonlethal Concepts: Implications for Air Force Intelligence," *Airpower Journal*, Vol. 8, No. 4, Winter 1994, pp. 26–33.

Parton, James, "The Thirty-One Year Gestation of the Independent USAF," *Aerospace Historian*, September 1987, pp. 151–152.

Perlez, Jane, "Aid Agencies Hope to Enlist Military Allies in the Future," *New York Times*, August 21, 1994, p. IV-6.

Pine, Art, "Deployments Take Toll on U.S. Military," *Los Angeles Times*, March 19, 1995, p. 1.

RAND, "Can the United States Increase Reliance on the Reserves?" *RAND Research Brief*, Santa Monica, Calif.: RAND, RB-7501, September 1994.

Roos, John G., "Help Humanity—Don't Hurt DoD," *Armed Forces Journal International*, September 1994.

Salerno, Steve, "The Cutting Edge of Combat Medicine," *The American Legion*, March 1995, p. 20.

Schmitt, Eric, "G.O.P. Military Power Assails Troop Readiness," *New York Times*, November 17, 1994, p. 22.

Schmitt, Eric, "Miliary's Growing Role in Relief Missions Prompts Concerns," *New York Times*, July 31, 1994, p. 3.

Setear, John K., et al., *The Army in a Changing World: The Role of Organizational Vision*, Santa Monica, Calif.: RAND, R-3882-A, 1990.

Serrano, Richard, "Military on Trial as Pilot Accused in Fatal Downing," *Los Angeles Times* (Washington), November 8, 1994, p. 1.

Summers, Harry G., Jr., "First Priorities for the Military," *Washington Times*, November 17, 1994, p. 18.

United States General Accounting Office, Operation Desert Storm, *Problems with Air Force Medical Readiness*, Washington, D.C., GAO/NSIAD-94-58, December 1993.

United States Government, *Basic Aerospace Doctrine of the United States Air Force*, Vol. 1, March 1992, pp. 259–261.

"War and Police," *Washington City Paper*, July 22, 1994, p. 17.

Watkins, Steven, "AWACS Crew Unlikely to Face Harsh Discipline," *Air Force Times*, August 29, 1994, p. 4.

Watkins, Steven, and Vago Muradian, "Pushing the Limits," *Air Force Times*, August 29, 1994, pp. 12, 16.

Weinschenk, Andrew, "In Rwanda's Wake Pentagon Draws Closer to U.N., Relief Groups," *Defense Week*, Vol. 15, No. 47, November 28, 1994, pp. 1, 12.

Zinni, Lieutenant General A. C., USMC. Remarks to the participants of EMERALD EXPRESS '95, Camp Pendleton, California, April 9–14, 1995.